Emulsions

Editor

Prof. Dr. A. WEISS – München

with 102 figures and 33 tables

Springer-Verlag Berlin Heidelberg GmbH 1978

ISBN 978-3-662-16093-0 ISBN 978-3-7985-1805-6 (eBook)
DOI 10.1007/978-3-7985-1805-6

CONTENTS

Dieser Ausgabe ist eine Mitteilung des Verlages beigefügt

PROGRESS IN COLLOID AND POLYMER SCIENCE

Fortschrittsberichte über Kolloide und Polymere

Supplements to "Colloid and Polymer Science" · Continuation of „Kolloid-Beihefte"

Vol. 63 1978

Progr. Colloid & Polymer Sci. 63, 1–9 (1978)

ISSN 0340-255 X

Department of Chemistry, University of Missouri, Rolla, Missouri (USA) and Swedish Institute for Surface Chemistry, Stockholm (Sweden)

Microemulsions in the water-potassium oleate-benzene system

S. Friberg and *I. Burasczenska*

With 7 figures and 1 table

(Received January 24, 1977)

Abstract

Microemulsion regions in the system water-potassium oleate-pentanol-benzene were determined by titration and equilibrium compositions in multiphase areas were analyzed.

The results showed the W/O microemulsions be inverse micellar solutions in accordance with earlier results. O/W microemulsions are related to the aqueous micellar solutions but with a more complex structure. At high electrolyte content the inverse micellar solution was in equilibrium with a dilute aqueous solution. The water solubilizing capacity of W/O microemulsions passed through a maximum with increasing soap/alcohol ratio, and excess water gave a phase transition to a lamellar liquid crystal at low electrolyte content of the water. The lamellar liquid crystalline phase showed no extremum of properties at compositions in equilibrium with W/O microemulsion of maximum water solubility and the basis for optimum stability of the inverse micellar solutions at a certain alcohol soap ratio was referred to conditions within the solution. The reason for this maximum solubilization capacity was found not to be related to intermicellar interactions but was confined to intra-micellar conditions. Of these conditions of the electric double layer were decisive.

Introduction

Technical applications of microemulsions have been known for a long time (1); the scientific literature observed the phenomenon more than thirty years ago (2). They are transparent systems with high content of both water and hydrocarbon and in contrast to normal emulsions they have been considered as thermodynamically stable (3).

The structure and the reason for the stability of microemulsions has been intensely discussed using mainly two approaches. The Schulman-Prince school (4–9) concentrated its efforts on the interfacial tension and its first moment stressing the necessity of a negative interfacial tension for the stability of the systems. This was obviously an oversimplification (10), *Adamson* (11), *Gillberg* (12), *Shinoda* (13) and *Ahmad* (14) pointed out the colloidal nature of the microemulsions. Recently *Ruckenstein* (15) has given a complete treatment demonstrating the relative importance for the stability of Van der Waals forces, interfacial tension, the repulsion forces from the electric double layer and the entropic contribution. All combinations gave a small but positive interfacial tension in agreement with independent calculations by *Reiss* (16).

Shah and coworkers (9, 17) have investigated

the inversion of a W/O microemulsion to an O/W system at addition of water and suggested the formation of several different intermediate structures of which the existence of a lamellar crystal has been directly verified by electron microscopy.

The possibility that the microemulsions may in fact not be thermodynamically stable but enjoy kinetic stability only has been brought to general attention by *Rosano* and *Gerbacia* in a series of articles (18, 19).

The questions of the microemulsion structure, stability and phase transitions have so far not received a final answer in spite of the rich literature on the subject. Since the phenomena are of considerable scientific interest and since the application of microemulsions in the form of low-tension systems in the tertiary oil recovery (20–25) is a highly significant technical development, we considered an investigation on the phase equilibria in a four component model system to be of general interest.

The system water (H_2O), potassium oleate, (KOl), pentanol (C_5OH) and benzene (C_6) was chosen to facilitate comparison with earlier investigations.

Experimental

Materials

The following chemicals were used without further purification. Oleic acid (cis-9-octadecanoic acid) Fluka AG 99%; 1-Pentanol (Fluka AG $\geq$99%); Benzene (Merck zur Analyse); Potassium (Riedel & Hahn AG), Sodium chloride (Merck p.a.) and twice distilled water. The analysis was made using radioactively labeled materials [9,10(N)-3H]Oleic acid 5 mCi/mole, (Radiomechanicals Centie Ltd) and [1,^{14}C] Pentanol, (ICN Pharmaceuticals Inc. Life Sci. Group).

Phase equilibria

The solubility areas were determined by titration until turbidity and the solubility limits checked by long time storage of samples on both sides of them. Separation of the phases was obtained by ultracentrifugation in a Beckman L2-65B followed by long time storage of the phases in contact until the compositions remained constant. For the liquid phases one month was found sufficient; when liquid crystalline phases were involved longer times were needed.

The composition of the samples was determined using ultraviolet spectroscopy to analyze the amount of benzene, Fischer titration to determine the amount of water and a Tri-Corb Packard 314A liquid scintillation apparatus to analyze the amounts of labeled pentanol and potassium oleate.

Results

The results will be presented in the following order. The regions for W/O microemulsions and the influence of electrolyte on them will first be shown. The equilibrium conditions of the water-pentanol-potassium oleate system and the changes brought forward by addition of electrolyte will follow and finally some features of the phase equilibria in the four component system will be treated.

W/O microemulsions

Figure 1 shows the W/O microemulsion areas at different levels of hydrocarbon. The base triangle (H_2O, KOl, C_5OH; fig. 1B)

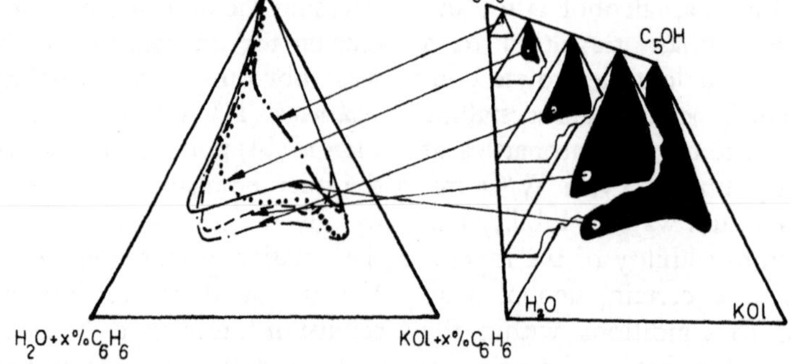

Fig. 1. W/O microemulsion regions are a direct continuation (*B*) of the inverse micellar area in the system of the three structure-forming elements water (H_2O), surfactant (potassium oleate, KOl) and cosurfactant (pentanol, C_5OH). The microemulsion regions, when projected onto this plane (*A*), remain approximately constant to hydrocarbon concentrations of 50%

shows the solubility area in the system of the three structure-forming elements. This area is, of course, no microemulsion since no hydrocarbon is present; it contains ion pairs and inverse micelles as has been indicated by *Ekwall* (26, 27). At higher levels of hydrocarbon the microemulsion areas were found as a *direct* continuation of the inverse micellar regions observed in the system with no hydrocarbon present and they contained identical structures (28); ion pairs and micelles. The presentation of microemulsion areas in this form illustrated the importance of the association conditions of the three structure-forming elements. The regularity in the solubility limits with varied hydrocarbon content was obvious (fig. 1).

One such regular variation was the *minimum* water content necessary to form a benzene/alcohol solution of water and soap. Figure 1 A demonstrates this minimum water content to be independent of the hydrocarbon content, related only to the *ratio* between soap and water. The water/soap molecular ratio was 3, which represents the minimum number of water molecules necessary to bring one molecule of soap into solution in pentanol or in the pentanol/benzene solutions.

The second regularity was the invariance of the *maximum* water solubility with hydrocarbon content when the amount of water was described as weight fraction of the three structure forming elements [$H_2O/(H_2O + KOl + C_5OH)$]. Figure 2 (X – X) shows this maximum water solubility to be independent of the benzene concentration up to 50 weight percent benzene. At higher benzene contents a reduction of the maximum water content was found; the curve in Figure 2 points toward zero for the weight fraction benzene equal to one. The soap/pentanol ratio for the composition containing maximum amount of water (fig. 1 A) varied with the benzene content. The lower curve in figure 2 shows a rise of the soap weight fraction at increasing benzene concentrations below 50% by weight followed by a reduction.

With these results an analogy between microemulsions and the alcohol solution of water and soap (26, 27) appeared justified. The W/O microemulsions should contain inverse micelles at high concentrations of water and ion pairs or small aggregates at lower water

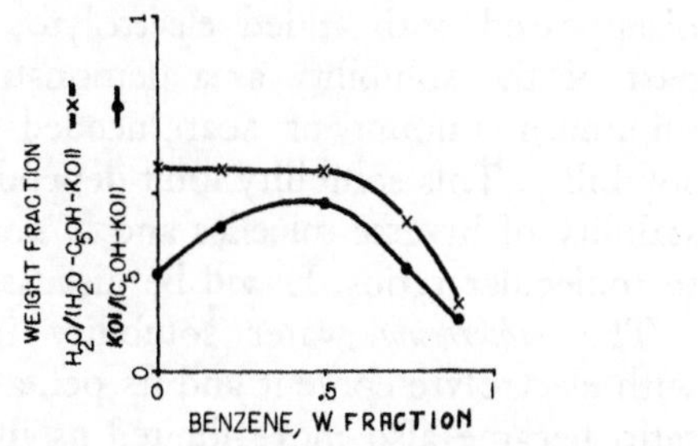

Fig. 2. The weight fraction of water ($H_2O/(H_2O + KOl + C_5OH)$, X – X) remained constant for hydrocarbon contents lower than 50% by weight. The soap content (o – o) in this maximum water composition increased to that hydrocarbon content and was reduced for higher hydrocarbon contents

concentrations which has been confirmed in an independent investigation (28). With this model a pronounced dependence of the solubility areas on the electrolyte content of the water could be expected considering also the importance of the osmotic pressure on the equilibrium conditions (11). Figure 3 shows the modifications of the solubility area, which were obtained, when sodium chloride was added to the water.

The *minimum* water content needed to obtain solubility was increased with the sodium chloride content as shown by the solubility limit to the right in the diagram. Calculations of the water/sodium chloride molar ratio for the limit gave a value 34 ± 3 with 95% confidence. The solubility of water in pentanol

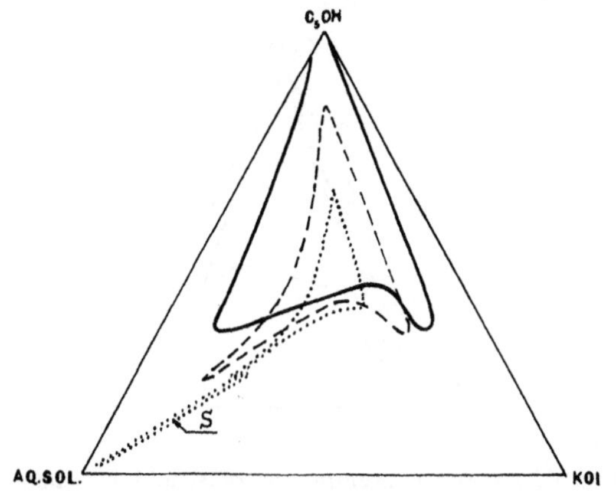

Fig. 3. Solubility areas for the three structure forming elements at different electrolyte content in the water (AQ. SOL.). ——— No electrolyte; - - - - 0.5M NaCl; .... 1M NaCl; (S explained in the text)

disappeared with added electrolyte; the left part of the solubility area demonstrates the minimum amount of soap needed to give solubility. This solubility limit depends on the stability of inverse micelles and is not related to molecular ratios. It will be discussed later.

The *maximum* water solubility increased with electrolyte content and its pentanol/soap ratio became also more limited as illustrated by the narrow solubility channels pointing towards the water corner. The soap/pentanol ratio for this channel was increased with increased electrolyte content. For the 1M NaCl solution the extremely narrow channel towards the water corner was discontinuous in the water range 59–64% and two solutions were observed. The more water rich solution did not reach the aqueous corner and was not an aqueous micellar solution. Its behavior is similar to the surfactant phase (29–31), which has such importance for the tertiary oil recovery technology (20–25).

These results on the variations in phase regions justified investigations on phase equilibria in these systems. The equilibrium conditions of the pertinent parts of the system of the three structure-forming elements will first be described followed by the microemulsion systems.

Phase equilibria H_2O, KOl, K_5OH

The results showed a two-phase area at low concentration of soap ($L_1 - L_2$, fig. 4). The aqueous solution of soap with solubilized alcohol was in equilibrium with a pentanol solution. Increased amount of soap in the latter gave a moderate increase of the solubility of water from 10 to 14 weight percent. At that water percentage the pentanol solution became part of a three-phase system in which the third phase was a lamellar liquid crystalline phase (*N*, fig. 4). For higher percentages of water, the pentanol solution (L_2) was in equilibrium with the lamellar liquid crystalline phase (*N*) only. This part of the pentanol solution region contained inverse micelles; (26–28) the water poor part of it in equilibrium with the aqueous micellar solution contained ion pairs or other such small aggregates. Hence solutions with inverse micelles were not found in equilibrium with an aqueous solution containing normal micelles.

The equilibria of soap and alcohol with an aqueous sodium chloride solution should be represented in a 3-dimensional diagram but, such a representation is difficult to read because the amounts of salt are small compared to the other components. The essentials of the phase equilibria are found in the projection onto the base plane of H_2O, KOl and C_5OH (fig. 5). Contrary to the equilibria involving water (fig. 4) the *entire* limit towards the water/soap axis of the inverse micellar solution to *maximum* water content was now in equilibrium with an aqueous solution of only small amounts of soap (0.08%) and pentanol (0.8%). This dilute aqueous solution was also a part of the three-phase area with the inverse micellar

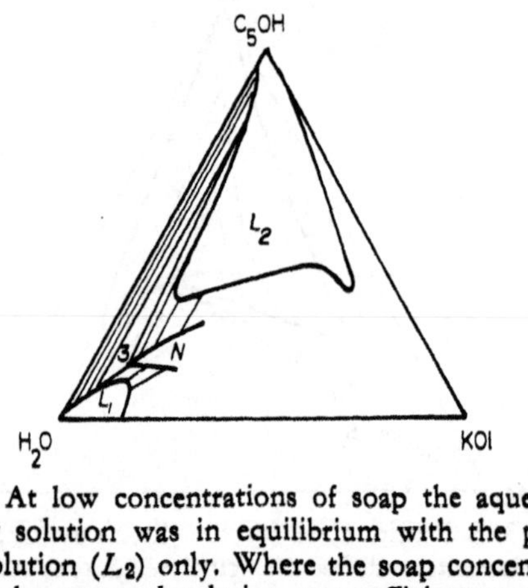

Fig. 4. At low concentrations of soap the aqueous micellar solution was in equilibrium with the pentanol solution (L_2) only. Where the soap concentration in the pentanol solution was sufficient to give inverse micelles the equilibrium was shifted to include the lamellar liquid crystalline phase (*N*)

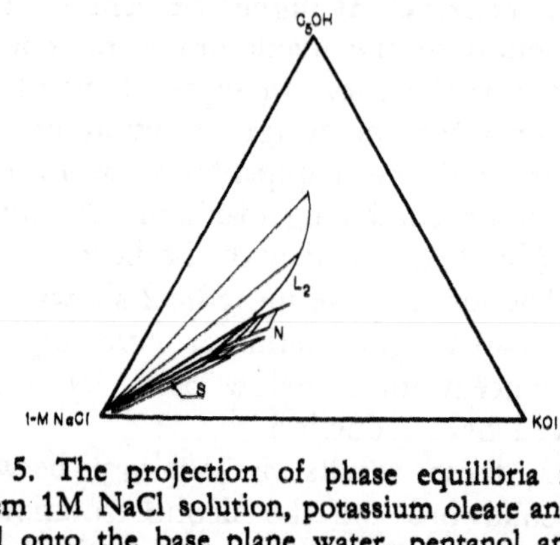

Fig. 5. The projection of phase equilibria in the system 1M NaCl solution, potassium oleate and pentanol onto the base plane water, pentanol and potassium oleate showed the entire left hand part of the inverse micellar solution to be in equilibrium with a dilute aqueous solution of soap and pentanol. The difference from Fig. 4 is obvious.

solution of maximum water content (L_2) and the lamellar liquid crystal (N). The difference of the inverse micellar solution region from the corresponding solubility area in figure 3 is due to partition of electrolyte between the inverse micellar solution, the aqueous solution and the liquid crystalline phase. The narrow solubility channel (S in figs. 3 and 5) was not in equilibrium with the inverse micellar solution; only with the liquid crystalline phase and the aqueous phase. The maximal amount of water in the liquid crystalline phase was considerably reduced by the electrolyte content.

Phase equilibria in the microemulsion system were restricted to compositions without NaCl in the 50% benzene plane. The basic objective was to estimate the concentration of soap in the inverse micellar solution at which a liquid crystal was formed. The results are given in table 1.

The table 1 shows that the inverse micellar solution, the W/O microemulsion, was in equilibrium with an aqueous solution only at small concentrations of soap; at a hydrocarbon phase soap content of 0.8% (sample 4) the equilibrium was with a liquid crystalline phase. For a total composition with 90% hydrocarbon, last sample, concentrated aqueous solutions of soap were found in equilibrium with dilute hydrocarbon solutions.

Discussion

The results verified the identity between W/O microemulsions and inverse micellar solutions in accordance with earlier results by *Adamson* (11). *Gillberg* (12), *Shinoda* (13) and *Ahmad* (14). We are aware that this is not in accordance with recently published suggestions (32), but consider figure 1 B sufficient experimental evidence to bring this discussion to an end.

The results also verified the results by *Shah* and coworkers (9, 17) on the existence of a lamellar liquid crystalline phase in the transition region between W/O and O/W microemulsions. The presence of a lamellar liquid crystal is obvious in the phase equilibria of the three structure forming elements (fig. 4) and table 1 gives evidence of its equilibrium with W/O microemulsions of even lower soap concentrations. On the other hand we failed to detect phases with cylindrical association aggregates indicated by *Shah* (9); further investigations on these phenomena are being undertaken. It is essential to realize that the lamellar liquid crystalline phase may be dispersed in the W/O or O/W microemulsions as pointed out by *Ahmad* (14). Such a dispersion has the appearance of an isotropic liquid and may not change its semblance for months. *Rosano* and *Gerbacia* (18, 19) have studied the question of long term colloidal stability of microemulsion systems relating it to the colloidal stability of inverse micelles. A continuation of these studies on dispersions (liposomes) of a lamellar liquid crystalline phase in the microemulsion would be interesting.

The present results necessitate a complementary treatment for the stability of W/O microemulsions to the one introduced by *Ruckenstein* (15). *Ruckenstein* gave a fundamental basis for a phase transition from W/O to O/W isotropic liquid systems, which is the case at low soap concentrations or for the technically important case with high electrolyte content. For the present systems without electrolyte the inverse micellar phase with a soap concentration as low as 0.8% was found in equilibrium with (table 1) a lamellar liquid crystalline phase.

The present experimental knowledge of

Table 1. Equilibrium compositions, microemulsion system

Sample	Phase	KOl	C_5OH	C_6	H_2O
		Weight percentage			
1	O	.1	24.2	74.3	1.4
	W	3.5	3.1	.1	93.0
2	O	.1	15.8	83.8	.8
	W	1.9	3.0	.7	93.3
3	O	.4	13.3	86.9	.7
	W	5.1	4.5	5.1	85.1
4	O	.8	16.9	81.0	1.2
	LC	8.6	7.2	10.9	72.3
5	O	12.8	17.5	60.1	14.8
	LC	26.9	13.0	18.7	40.8
6	O	12.8	11.0	5.5	10.1
	LC	23.3	11.3	26.6	40.0
7	O	.1	.5	99.6	.2
	W	25.7	.7	.1	74.6

O = hydrocarbon phase
W = aqueous phase
LC = liquid crystalline phase

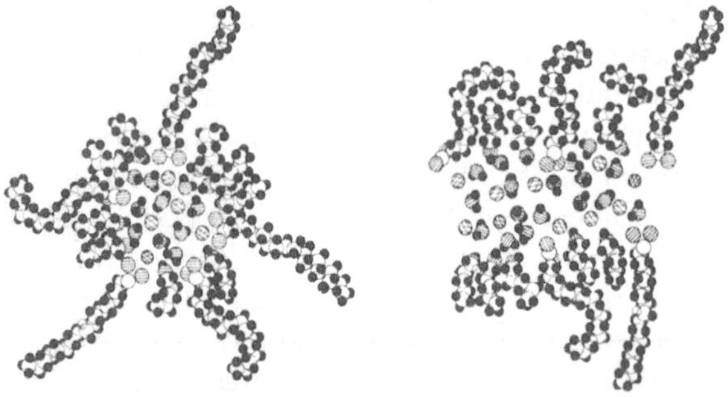

Fig. 6. The inverse micelles in the W/O microemulsions (left) were transformed to a lamellar liquid crystalline phase (right). ○, Carbon; ●, Hydrogen; ◍, Oxygen; ⊛, Potassium

microemulsion systems is not sufficient for a complete treatment of the transition from an inverse micellar phase to a lamellar liquid crystal (fig. 6). The better known system of the three structure forming elements (fig. 4) may, however, be of value for a later clarification of the more complex microemulsion systems. The phase transition in the system of water, potassium oleate and pentanol involved an inverse micellar solution and a lamellar liquid crystalline phase (L_2 and N, fig. 4). These phases had the following characteristics (fig. 6). The inverse micellar solution contained soap molecules in inverse micelles and in smaller aggregates (26, 27). The size of the inverse micelles is not known but could be estimated to lie in the radius range 20–50 Å (water core), using reasonable values (30–60 Å^2) for the area per molecule of soap and alcohol at the interface, for their ratio ($C_5OH/KOl = 6$–10), and for the amount of water in the micelles (25–50 percent by weight). The lamellar liquid crystal showed a layer spacing of 79.4 Å according to low angle X-ray diffraction data and it had a density of 0.99 g/cm^3 as determined by weighing known volumes of it. A formal calculation of the thickness of the amphiphilic layer (33) gave the value 23.8 Å and an average area per molecule of 53.4 Å^2. Such a model was, however, not internally consistent. A lamellar structure with all the pentanol molecules localized with the polar group close to the aqueous layer would leave voids in the hydrocarbon lamella and a modification was necessary. Modification included water molecules between the polar molecules which is reasonable considering the high solubility of water in pentanol and *vice versa*. An average of three molecules of water per amphiphile gave the layer a thickness of 32.6 Å and an area per amphiphile of 39.0 Å^2, which is acceptable. With these estimations the association structures of the two phases appear as in figure 6. They form the basis for the following discussion.

The phase transition to a lamellar phase was decisive for the maximum water solubility in the inverse micellar phase at a pentanol/soap ratio of 10 (fig. 4). The conditions that lead to the reduction of water solubility at soap/alcohol ratios in excess of this ratio could depend on changes in the lamellar liquid crystal or in the isotropic solution. The lamellar phase showed no structural changes at the composition in equilibrium with the inverse micellar solution at optimum water content and the structural reasons for the maximum must be localized within the inverse micellar solution. The choice is between variation in micelle number or micelle size as the decisive factor. The former mechanism leads to a consideration of intermicellar interactions as the responsible factor; the second one puts emphasis on conditions within each micelle.

The present results are in favor of micellar size being the decisive factor against the initial invariance with hydrocarbon content of the water weight fraction [$H_2O/(H_2O + KOl + C_5OH)$] for maximum solubility of water (fig. 2). Added hydrocarbon would be dissolved in the pentanol and hence change the intermicellar distance. The interparticle distance/particle radius ratio will be

$$d/r = (1.81/\sqrt[3]{\phi}) - 2 \qquad [1]$$

assuming hexagonal close-packing of spherical particles occupying a space fraction ϕ. A change of ϕ from 0.5 (no hydrocarbon) to 0.25 (50% hydrocarbon) increases d/r with a factor of three. It is difficult to envisage particle/particle interactions that would not be influenced by an increase of interparticle distance/particle radius ratio of that magnitude. Hence the attention may be concentrated towards the different factors, which stabilize an inverse micelle (34–36).

The stability factors for water-free inverse micelles have been treated by *Eicke* (35) who has given the following factors.

1. Solvent-solvent dispersion interactions.
2. Dispersion interactions between the hydrocarbon chains in the micelle and the solvent.
3. Dispersion interactions between the hydrocarbon groups in the micelle.
4. Coulombic interactions in the form of a) Repulsion terms between positive/positive and negative/negative charges. b) Corresponding attraction terms. c) Electrostatic energy between positive and negative charges in the monomers.

The electrostatic energy for an inner electric double layer in an inverse micelle containing water has been treated by *Levine* (3), who gave the free energy for the formation of a diffuse inner double layer with a Gouy-Poisson distribution.

The present results form a sufficient basis for exclusion of some of the factors. The influence of factors 1 and 2 may be excluded against the invariance of water solubility with addition of hydrocarbon (fig. 2, X – X). The two remaining factors; 3. dispersion interaction between hydrocarbon chains within the micelle and 4. Coulombic forces, both had an influence.

The dispersion interaction influence has been illustrated by the variation of the maximal solubility of water with alcohol chain length in inverse micellar solutions, which contain an identical soap, a sodium caprylate (27). The series nonanol-pentanol has given the following maximum water contents 33% (C_9OH); 33% (C_8OH); 34% (C_7OHO); 38% (C_6OH) and 62% (C_5OH).

This enhanced water solubilization with shorter chain length of the alcohol is understood against the phase transition between the inverse micellar phase and the lamellar liquid crystalline phase (fig. 6). The possibility that a shorter chain length of the alcohol should give enhanced stability to the inverse micelle appears small (fig. 6, A) and an explanation involving a *destabilization* of the liquid crystalline phase (fig. 6, B) appears more reasonable. The enhanced chain length difference could be expected to give an increased disorder (37, 38) and as a consequence destabilization of the liquid crystalline phase. The reduced stability of the liquid crystalline phase leads to higher percentage of water in the inverse micellar solution before a phase transition takes place.

It is important to realize that this mechanism is decisive for the variation of maximal solubility of water in the inverse micellar phase with changed alcohol chain length.

For the variation of water solubilization and its maximum with soap/alcohol ratio at a certain alcohol chain length in the present investigation the Coulombic forces must also be taken into account. In fact the constant soap/alcohol ratio at maximum water content in inverse micelles for different alcohols indicates the Coulombic forces to be dominant.

No direct information is available on the conditions in the inner electric double layer of an inverse micelle (34); indirect information may be obtained from the results on counter ion binding by NMR (39–41). The relaxation rates for Na^+ ions in the system sodium octanoate, pentanol and water has been investigated by *Rosenholm* and *Lindman* (41). Plotting their results for the inverse micellar solution gave a minimum for the counter ion binding at the maximum water content of the inverse micellar solution (fig. 7). The corresponding values for the lamellar liquid crystalline phase showed no extremum at composition in equilibrium with the inverse micellar solution.

Since variation of the chain length of the alcohol did influence the maximum water solubilization capacity, but not the soap/alcohol ratio at which the maximum was found, a reasonable conclusion is that the conditions of the diffuse electric double layer is the decisive factor for the obtained maximum

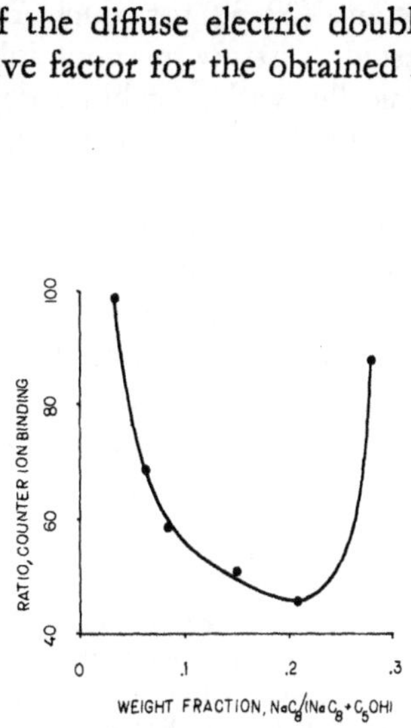

Fig. 7. The counter ion binding to the surfactant reached a minimum at maximum water content of the inverse micelle (41)

of water solubilization versus soap/alcohol ratio.

This conclusion is supported by the results from the present systems containing electrolytes. Addition of electrolyte to the water enhanced the dependence of the maximum water solubility on the soap/alcohol ratio (fig. 3) and gave a higher soap/alcohol ratio for the maximum, increased the maximum water solubility.

These changes may be expected against the present views on the importance of the phase transition to a lamellar liquid crystal for the stability of the inverse micelles. The maximal stability of the lamellar liquid crystalline phase at high water contents has earlier (42) decisively been shown to depend on the electric surface potential; which means a dependence on the double layer characteristics. Addition of electrolyte gives a compression of the electric double layer and hence a reduced stability of the lamellar phase. The changes of water solubilization (figs. 3, 5) in the present investigation did follow an expected pattern; the maximal water content of the liquid crystal was reduced and the region of inverse micellar solution correspondingly enhanced.

At high electrolyte concentrations a water-rich isotropic liquid phase (S. fig. 3) was observed. In spite of its high water content it could not be infinitely diluted with water and may not be called an aqueous phase. It is similar to microemulsions containing nonionic surfactants (29, 30) for which *Robbins* (43) has suggested a structure containing large spherical micelles with a stability dependence entirely on the bending component of the interfacial energy. The sectorial shape of the solubility region is in itself a support of such a structure, but the fact that the solubility region does not reach 100 % water demonstrates other factors also to be involved. Since this structure is of paramount importance in the tertiary oil recovery (20–25) the full clarification of it may be expected within shortly.

Summary

W/O microemulsions were shown to be identical to the inverse micellar solutions found in systems of alcohol with dissolved water and surfactant. The stability of the inverse micellar system was determined by the conditions in the inner electric double layer, the phase transition being to a lamellar liquid crystalline phase instead of an aqueous solution.

Addition of electrolyte increased the maximum amount of dissolved water in the inverse micellar solution and required a higher soap/alcohol ratio to give this solubility. This result was expected from the compression of the electric double layer giving reduced stability to the lamellar liquid crystalline phase with high water content. The increase of maximal water content of the inverse micellar solution arose from a delay of the phase transition to a lamellar liquid crystalline phase with increasing water content.

Zusammenfassung

Im System Wasser-Kaliumoleat-Pentanol-Benzol wurden die Mikroemulsionsbereiche durch Titration bestimmt und die Gleichgewichtszusammensetzungen in Mehrphasengebieten analysiert.

Die Ergebnisse zeigten, daß W/O-Mikroemulsionen inverse mizellare Lösungen sind. O/W-Mikroemulsionen sind verwandt mit wäßrigen mizellaren Lösungen, haben aber komplexere Struktur. Bei hohen Elektrolytgehalten waren die inversen mizellaren Lösungen im Gleichgewicht mit einer verdünnten wäßrigen Lösung. Die Solubilisierungskapazität von W/O-Mikroemulsionen für Wasser ging durch ein Maximum mit steigendem Verhältnis Seife/Alkohol. Überschüssiges Wasser gab einen Phasenübergang zu einem lamellaren Flüssigkeitskristall bei niedrigen Elektrolytgehalten im Wasser.

References

1) *Kokatnur, V. R.*, U.S. Patent 2, 111, 100 (1935).

2) *Hoar, T. P., H. H. Schulman*, Nature (London) **152**, 102 (1943).

3) *Schulman, J. H., D. P. Riley*, J. Colloid Sci. **3**, 383 (1948).

4) *Bowcott, J. E. L., J. H. Schulman*, Z. Electrochem. **59**, 283 (1955).

5) *Stockenius, W., J. H. Schulman, L. Prince*, Kolloid-Z. **169**, 170 (1960).

6) *Schulman, J. H., J. B. Montagne*, Ann. N.Y. Acad. Sci. **92**, 3661 (1961).

7) *Prince, L. M.*, J. Colloid Interface Sci. **23**, 165 (1967).

8) *Rosano, H. L., R. C. Peiser, A. Eydt*, Rev. Tran. Corps. Gras. **16**, 249 (1969).

9) *Shah, D. O., R. M. Hamlin*, Science **171**, 483 (1971).

10) *Miller, C. A., L. E. Sciven*, J. Colloid Interface Sci. **33**, 360 (1970).

11) *Adamson, A. W.*, Ibid. **29**, 261 (1969).

12) *Gillberg, G., H. Lehtinen, S. Friberg*, Ibid. **33**, 40 (1970).

13) *Shinoda, K., H. Kunieda*, Ibid. **42**, 381 (1973).

14) *Ahmed, S. I., K. Shinoda, S. Friberg*, Ibid. **47**, 32 (1974).

15) *Ruckenstein, E., J. C. Chi*, J. Chem. Soc. Faraday Trans. II, **71**, 1690 (1975).

16) *Reiss, H.*, J. Colloid Interface Sci. **53**, 61 (1975).

17) *Falco, J. W., R. D. Walker, Jr., D. O. Shah*, Am. Ind. Chem. Engg. **20**, 510 (1974).

18) *Gerbacia, W. H. L. Rosano* J. Colloid Interface Sci. **44**, 242 (1973).

19) *Rosano, H. L.*, J. Soc. Cosmetic Chem. **25** 609 (1974).

20) *Foster, W. R.*, J. Petroleum Tech. **25**, 205 (1973).

21) *Healy, R. N., R. L. Reed*, Soc. Petr. Eng. J. **14**, 491 (1974).

22) *Healy, R. N., R. L. Reed, D. G. Stenmark*, Soc. Petr. Eng. Preprint 5565 (1975).

23) *Wilson, P. M., Murphy, C. L., W. R. Foster*, Ibid. 5812 (1976).

24) *Miller, C. A. R.-N. Hwan W. J. Burton, T. Fort Jr.*, J. Colloid Interface Sci.

25) *Ruckenstein, E.*, Proc. NERM, ACS Meeting (Albany 1976).

26) *Mandell, L., P. Ekwall*, Acta Polyt. Scand. **74**, 1 (1968).

27) *Ekwall, P., L. Mandell*, Acta Chem. Scand. **21**, 1612 (1967).

28) *Sjoblom, E.*, To be published.

29) *Shinoda, K., T. Ogawa*, J. Colloid Interface Sci. **24** 56 (1967).

30) *Frank, S. G., G. J. Zografi*, J. Colloid Interface Sci. **29**, 27 (1969).

31) *Friberg, S., I. Lapczynska*, Progr. Colloid & Polymer Sci. **56**, 16 (1975).

32) *Prince, L.*, J. Colloid Interface Sci. **52**, 182 (1975).

33) *Fontell, K., P. Ekwall*, Acta Polyt. Scand. **74**, 86 (1968).

34) *Levine, S., K. Robertson* J. Phys. Chem. **76**, 876 (1972).

35) *Eicke, H. F., H. Christen*, J. Colloid Interface Sci. **46**, 417 (1974).

36) Idem, Ibid. **48**, 281 (1974).

37) *Charvolin, J.*, In: Lyotropic Liquid Crystals (Ed. S. Friberg), ACS Advances in Chemistry Series No. 152, p. 101 (1976).

38) *Reeves, L. W.* and collaborators (To be published).

39) *Lindblom, G., B. Lindman*, J. Phys. Chem. **77**, 2531 (1973).

40) *Gustavsson, B. Lindman*, J. Am. Chem. Soc. **97**, 3923 (1975).

41) *Rosenholm, J. B., B. Lindman*, To be published.

42) *Krog, N., J. B. Lauridsen* in Food Emulsions (Ed. *S. Friberg*) p. 104 (New York 1976).

43) *Robbins, M.*, The Theory of Microemulsions, A. I. Ch. E. Nat. Meeting, Tulsa, Oklahoma, March 1974.

Author's address:

Professor *Stig Friberg*
Department of Chemistry
University of Missouri – Rolla
Rolla, Missouri 65401
(USA)

Progr. Colloid & Polymer Sci. **63**, 10–14 (1978)

ISSN 0340-255 X

Department of Food Science and Nutrition, Queen Elizabeth College, University of London, London (England)

Mechanism of temperature induced phase inversion in O/W emulsions stabilised by O/W and W/O emulsifier blends

P. Sherman and *C. Parkinson*

With 2 figures

(Received July 7, 1976)

Introduction

The solubilization of hydrocarbons in aqueous solutions of non-ionic emulsifiers, and of water in non-aqueous solutions of non-ionic emulsifiers, has been actively studied by *Shinoda* and co-workers (1–6) over many years. They showed that solubilization is very dependent on temperature and that it changes drastically at a certain temperature in any given system. Furthermore, at this temperature there is a drastic change in the stability of emulsions prepared from the same components as those utilized in the solubilization studies. This temperature was named the phase inversion temperature (PIT) because a slight increase for an O/W emulsion or decrease for a W/O emulsion resulted in emulsion inversion.

Shinoda (7) explained temperature controlled phase inversion on the basis of the phase diagram for the three component system, water, oil and non-ionic emulsifier, and the effect of temperature thereon. In the phase diagram studies only one emulsifier was used and its characteristics changed from hydrophilic to lipophilic with increasing temperature. Many practical O/W emulsions incorporate an O/W emulsifier (high HLB) with a W/O emulsifier (low HLB) so as to achieve a stability which is superior to that offered by the O/W emulsifier on its own. The present investigation was carried out to determine whether *Shinoda*'s explanation is applicable in this latter situation.

Experimental

Preparation of emulsions

O/W emulsions (primary emulsions) were prepared using Groundnut oil, distilled water and the two emulsifiers Tween 80 (Honeywill-Atlas) and Dimodan LS (Grinstead Products). Dimodan LS is a distilled monoglyceride prepared from refined Sunflower oil and it contains about 62% linoleic acid; Tween 80 is polyoxyethylene sorbitan monoleate. The O/W emulsions contained 49.58% (wt/wt) Groundnut oil, 48.42% (wt/wt) distilled water and 2.0% (wt/wt) combined emulsifiers. In one emulsion series (series 1) 0.42% Dimodan LS was used in conjunction with 1.58% Tween 80; in the second series (series 2) 1.025% Dimodan LS was used in conjunction with 0.975% Tween 80. These proportions correspond to Dimodan: Tween 80 1:4 and 1:1 molecular ratios respectively.

The Dimodan LS was dissolved in the Groundnut oil and the Tween 80 in the distilled water. Both emulsifiers dissolved readily in their respective liquid phases. The oil phase was then added slowly to the aqueous phase while stirring vigorously with a glass rod. Finally, each mix was homogenized for 10 secs with a Silverson mixer-homogenizer set at the lowest speed using the mixer element with the smallest mesh size.

A few emulsions were prepared in which the emulsifier blend was a 1:1 ratio (wt/wt) of Span 20 and Tween 20, the emulsifier blend still forming 2.0% (wt/wt) of the total emulsion.

Phase inversion temperature

The PIT values of freshly prepared emulsions were evaluated with a Du Pont 900 Thermal Analyzer by the differential thermal analysis technique described elsewhere (8).

Microscopic and photographic examination of emulsions

Emulsion samples were heated in loosely stoppered glass test tubes in a water bath at a rate of approximately 1 °C min^{-1}. The samples were stirred gently at frequent intervals.

Small samples of each emulsion were withdrawn at regular intervals for microscopic examination. For this purpose 0.1 ml emulsion was diluted with 10 ml aqueous phase. The slide and coverslip were incubated at the same temperature to which the emulsion had been heated when the sample was taken from the test tube. Photographs of the diluted emulsion samples were taken with a Zeiss Photomicrograph 1 microscope using Ilford FP4 film. The total magnification under these conditions was 3,200.

Results

Phase inversion temperature

The PIT's of series 1 and series 2 emulsions were 90.0 °C and 77.0 °C respectively.

Photomicrographic appearance of emulsions during heating

Figures 1 and 2 show how the internal structure of series 1 and series 2 emulsions respectively alter as the temperature is raised.

In both series a multiphase (W/O/W) emulsion structure develops due to the appearance of water drops within the oil drops. The size of the oil drops increases with increasing temperature, but the water drops are

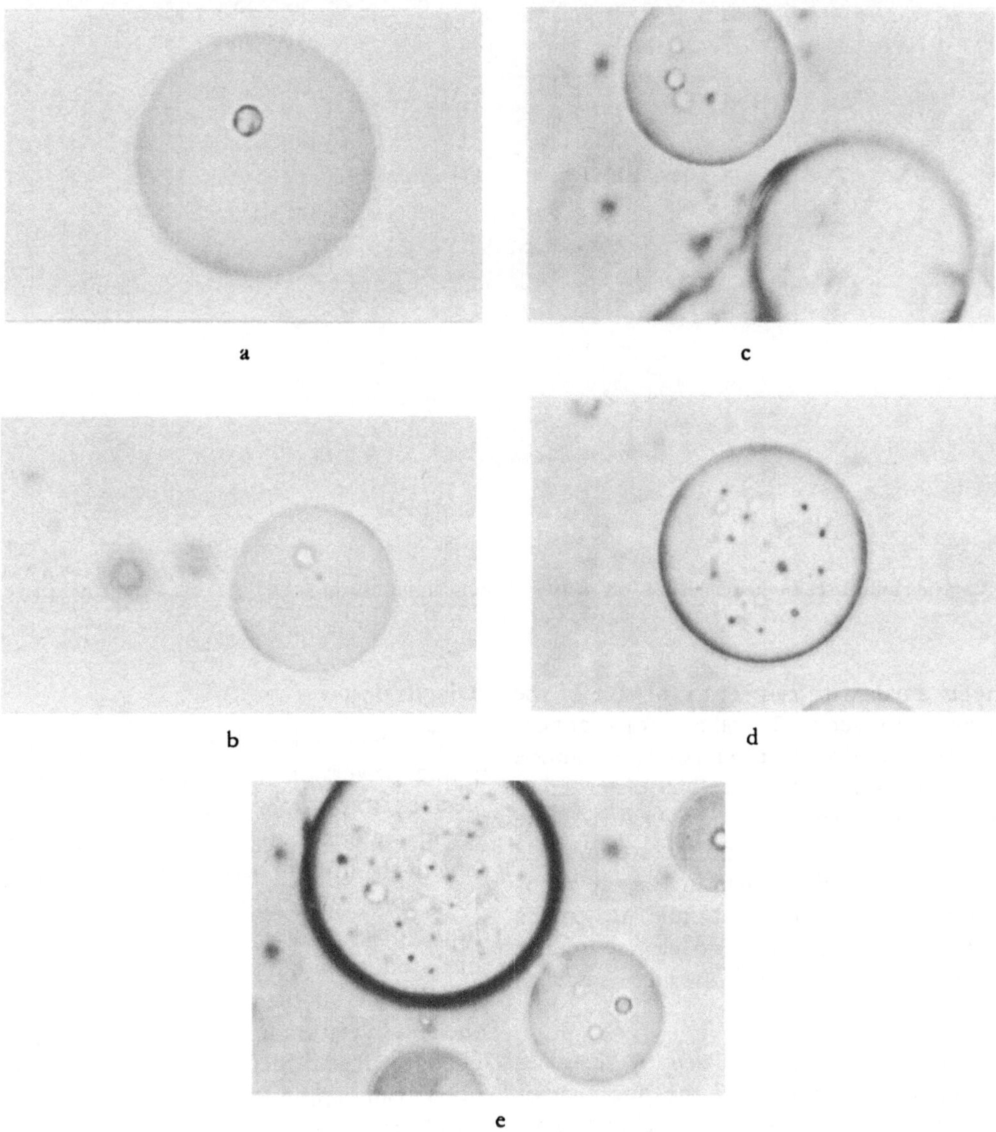

Fig. 1. The influence of temperature on series 1 emulsions during D.T.A. measurement of P.I.T. a) 45 °C, b) 55 °C, c) 65 °C, d) 75 °C, e) 85 °C

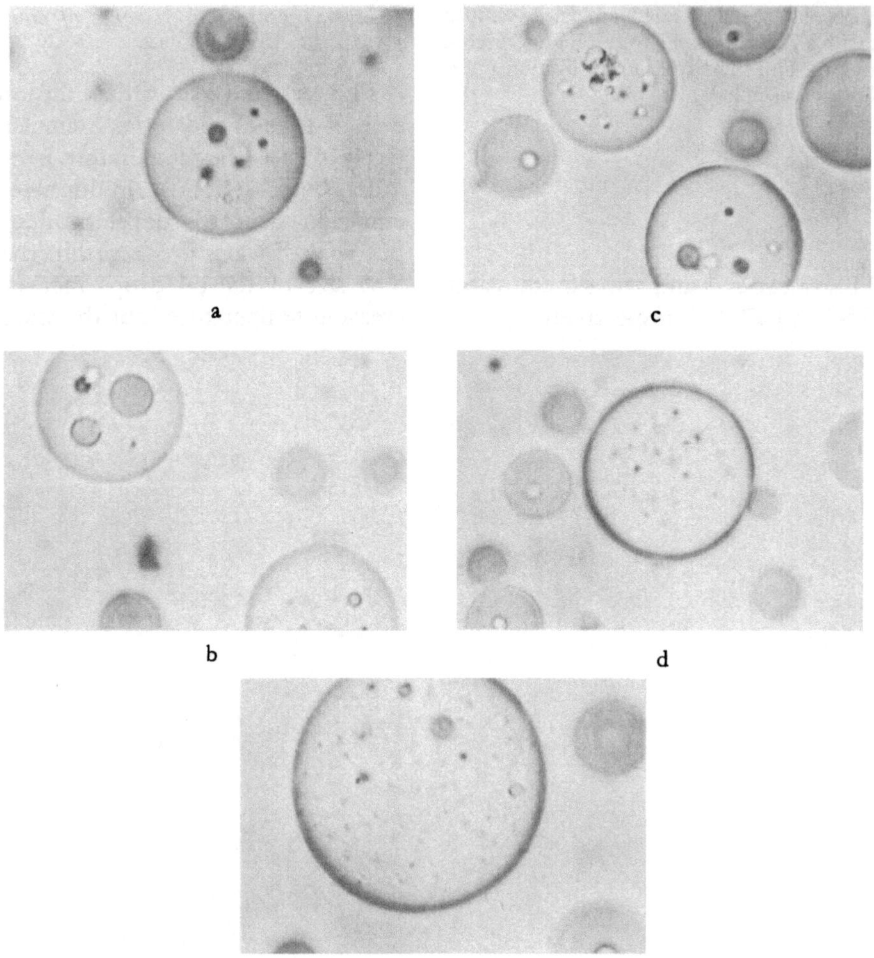

Fig. 2. The influence of temperature on series 2 emulsions during D.T.A. measurement of P.I.T. a) 35 °C, b) 45 °C, c) 55 °C, d) 65 °C, e) 75 °C

relatively small with respect to the size of the oil drops. In series 1 water drops appear within the oil drops at temperatures above 65 °C but in series 2 they are first observed at the much lower temperature of 45 °C. As the temperature increases beyond these points the number of water drops within the oil drops increases, and this trend continues up to the respective PIT's. There is no clear evidence that the size of the water drops increases with increasing temperature.

The few emulsions stabilized by Span 20 and Tween 20 first showed the presence of water drops within the oil drops at about 45 °C. Above this temperature the number of water drops per oil drop increased as for series 1 and 2 emulsions.

Discussion

The mechanism for temperature induced phase inversion proposed by *Shinoda* (7) derives from solubilization studies on mixtures of oil, water and non-ionic emulsifier in which one of the liquid phases is progressively added to a solution of the emulsifier in the other liquid phase. The emphasis is on solubilization of the added liquid within micelles of the emulsifier located in the other liquid phase and how the solubilization is influenced by temperature changes. It is difficult to see how this mechanism can apply within a practical O/W emulsion system. In this latter situation there is no free oil that can be solubilized within emulsifier molecule micelles in the aqueous phase as the temperature is increased. The oil is dispersed as microscopic size drops which are stabilized by an adsorbed emulsifier film and although the rate of drop coalescence may increase as the temperature increases towards the PIT the oil within the drops is not released into the aqueous phase.

The ΔT data in DTA determinations of PIT (8) suggest that there is a steady continuous change in O/W emulsions as the temperature increases up to the PIT, at which point there is an abrupt phase change. If micellar solubilization was responsible for this effect one would expect the increase in solubilization as temperature increases to follow a similar pattern to the ΔT changes. Examination of the data published by *Shinoda* and his collaborators indicates that this is not the case and that the greatest change in micellar solubilization occurs near to the PIT.

Microscopic examination reveals that as the temperature of the O/W emulsions stabilised by O/W and W/O emulsifier blends is increased W/O/W emulsions of increasing complexity develops, and at the PIT they invert to a W/O emulsion. This sequence of events accords with the evidence offered by the DTA ΔT data. However, the origin of the water which appears in the oil drops requires explanation. The most likely source would appear to be the polyoxyethylene derivative Tween 80. Part of this emulsifier is adsorbed around the oil drops in the initially formed O/W emulsions as a complex with Dimodan LS molecules (9, 10) while the remainder forms micelles in the aqueous medium. The polyoxyethylene groups in the Tween 80 molecules have a relatively large hydration capacity, which arises from two sources. One of these is the hydrogen bonding of water molecules to ether oxygens of the polyoxyethylene chains, and the other is the physical entrapment of water within these chains which have a coil configuration (11–14). Light scattering, vapour pressure and viscosity data suggest that the water bonded to the ether oxygens decreases as the temperature rises whereas the water held within the polyoxyethylene coils increases (15). The bonding of water to the ether oxygens is believed to be mainly responsible for the solubility of polyoxyethylene derivatives in water.

It is difficult to conceive that water released from the ether oxygens of polyoxyethylene molecules within the micelles located in the aqueous phase reappears as water drops within the primary O/W emulsion oil drops. If micelles become less water soluble with increasing temperature as water bonded to the ether oxygen is lost the micelles containing physically entrapped water will, nevertheless, not be able to transfer into the oil drops via the interfacial layer because of their size, which actually increases with temperature (15). The only likely source of the water appearing within the oil drops is the Tween 80 molecules adsorbed around the oil drops in the primary O/W emulsion. Their hydrophilic character decreases as the temperature increases so that they gradually desorb and pass into the oil drops. There it is necessary for them to assume a new configuration and this releases the physically entrapped water which is emulsified and stabilised by Dimodan LS molecules present in the oil phase.

The decrease in the hydrophilic character of the Tween 80 molecules with increasing temperature will also alter the nature of the Tween 80 – Dimodan LS molecular complex (9, 10) around the oil drops in the primary O/W emulsion. As Tween 80 molecules move into the oil phase they will not be replaced by Dimodan LS molecules as adsorption at interfaces decreases with increasing temperature. Consequently, the interfacial layer may, in effect, become a semi-permeable membrane through which water can pass from the aqueous phase into the oil drops. This second mode of water transfer will become operative at a higher temperature than the mechanism described in the previous paragraph.

Summary

The temperature induced phase inversion of Groundnut oil-in-water emulsions stabilised by blends of Tween 80 and Dimodan LS is associated with the development of multiple phase emulsion drops (W/O/W emulsion). Drops become progressively more complex in structure as the temperature rises until eventually the emulsions invert to W/O. These changes are attributed to two effects which arise from the decrease in the hydrophilic character of the Tween 80 molecules adsorbed with Dimodan LS molecules around oil drops in the initially formed O/W emulsion. The Tween 80 molecules move into the oil drops where configurational changes release water originally entrapped in the polyoxyethylene coils. This results in modifications to the interfacial layer at higher temperatures so that aqueous phase can permeate through it and into the oil drops.

References

1) *Shinoda, K., H. Arai,* J. Phys. Chem. **68,** 3485 (1964).
2) *Saito, H., K. Shinoda,* J. Colloid Interf. Sci. **24,** 10 (1967).
3) *Shinoda, K., T. Ogawa,* J. Colloid Interf. Sci. **24,** 56 (1967).
4) *Shinoda, K., H. Saito,* J. Colloid Interf. Sci. **26,** 70 (1968).
5) *Shinoda, K., H. Saito,* J. Colloid Interf. Sci. **30,** 258 (1969).
6) *Saito, H., K. Shinoda,* J. Colloid Interf. Sci. **32,** 647 (1970).
7) *Shinoda, K.,* J. Colloid Interf. Sci. **24,** 4 (1967).
8) *Matsumoto, S., P. Sherman,* J. Colloid Interf. Sci. **33,** 294 (1970).
9) *Boyd, J. V., C. Parkinson, P. Sherman,* J. Colloid Interf. Sci. **41,** 359 (1972).
10) *Boyd, J. V., N. Krog, P. Sherman,* Symposium: Theory and Practice of Emulsion Technology p. 99 (London 1974).
11) *Elworthy, P. H., C. B. MacFarlane,* J. Chem. Soc., 311 (1964).
12) *Elworthy, P. H., C. McDonald,* Kolloid-Z. **195,** 16, (1964).
13) *Schick, M. J.,* J. Am. Oil Chemists Soc. **40,** 680 (1963).
14) *Lange, H.,* Kolloid-Z. **201,** 131 (1965).
15) *Nakagawa, T.,* In: Non-Ionic Surfactants, p. 572 (New York 1967).

Authors' address:

P. Sherman and *C. Parkinson*
Department of Food Science and Nutrition,
Queen Elizabeth College,
University of London,
Campden Hill Road,
London W8 7AH
England

Progr. Colloid & Polymer Sci. **63**, 15–26 (1978)

ISSN 0340-255 X

Pharmacy Department University of Aston in Birmingham (England)

The properties of O/W emulsions stabilised with a non-ionic surfactant and surfactant-polymer mixtures

*R. Buscall**)

With 12 figures and 2 tables

(Received August 1, 1975)

Introduction

The process of emulsion breakdown is complex and may involve several stages. For example, emulsion droplets stabilised with ionic surfactants can undergo flocculation (aggregation in the secondary minimum), coagulation (aggregation in the primary minimum) and finally rupture of the thin film separating the droplets (coalescence), whereas emulsions stabilised with polymers and non-ionic surfactants will undergo flocculation and coalescence. Most of the work on non-ionic surfactants as stabilisers has involved polyoxyethylene glycol derivatives and similar surfactants (1). However studies on thin liquid films have shown that Newton-black (primary minimum) films (2) can be very stable and implies that emulsions stabilised with simple, low-molecular weight non-ionic surfactants might have appreciable stability to coalescence. The first part of this paper deals with an investigation into the breakdown of emulsions stabilised by a simple non-ionic surfactant, decyldimethylphosphine oxide. The effect of electrolytes is also considered.

Surfactants are often used to improve emulsification when polymers are used to stabilise emulsions. Some surfactants have been shown to form association complexes with water-soluble polymers and these are found to be effective emulsion and dispersion stabilisers (3, 4). The second part of this paper is concerned with the interaction of decyldimethylphosphine oxide with poly(vinylalcohol) and their effect upon emulsification and emulsion stability.

Experimental

Materials

The sample of decyldimethyl-phosphine oxide was a gift from Professor *R. H. Ottewill.* From surface tension measurements the critical micelle concentration was found to be 3.8×10^{-3} mol dm^{-3}, sodium decyl sulphate, sodium dodecyl sulphate and cetyltrimethylammonium bromide were purified commercial products. The poly(vinylalcohol) used (ex. BDH Ltd) had a molar mass of 125 kg mol^{-1} and was approximately 80% hydrolysed. Light liquid paraffin B.P. (ex. Fisons Ltd) was passed through activated alumina columns prior to use and had an oil/water interfacial tension always in excess of 50 mNm^{-1}. The inorganic salts used were 'Analar' grade where this was available. Fresh, double distilled water was used throughout.

Methods

Interfacial Tensions

The pendant drop method was used for decyldimethylphosphine oxide solutions containing no polymer. Interfacial tensions were calculated with the aid of *Niederhauser* and *Bartell's* tables (5). All other measurements were made using a roughened platinum Wilhemy plate suspended from a recording microbalance (ex C.I. Electronics Ltd).

Microelectrophoresis

Mobility measurements were made using a commercial apparatus (Rank Bros) which had a Van Gils cell and employed direct illumination. The oil droplets studied had diameters in the range 1–10 μ and so the Smoluchowski equation was used to calculate zeta-potential values.

---

*) Present address: School of Chemistry University of Bristol.

Centrifugation

Emulsions containing 40% v/v oil were spun at speeds up to 18,000 r.p.m. for thirty minutes. After which the amount of oil liberated was estimated using a travelling microscope.

Kinetics

Emulsions containing 10% v/v oil were prepared by mechanical agitation or ultrasonic irradiation and typically had an initial model droplet diameter of ca. 3–4 μ. At appropriate time intervals the emulsions were mixed by gentle inversion and a 1 ml sample withdrawn. This was diluted with surfactant solution and examined microscopically. Mean droplet volume diameters were obtained by analysing several photomicrographs with the aid of a Zeiss TGZ 3 particle size analyser.

Surface Rheology

Measurements of the in phase component of the surface shear modulus, $G's(\omega)$ were made at a frequency of 4 Hz using a surface rheometer kindly made available by Drs. *Warburton* and *Sherriff*. This apparatus has been described in ref. 6.

Emulsification studies

20 mls of liquid paraffin and 30 mls of aqueous solution were coarsely emulsified by shaking and then subjected to ultrasonic radiation for one minute using a 'Soniprobe' type 7530 A. Care was taken to ensure the same probe and container geometry for each sample. Size distributions were determined from photomicrographs.

Results and discussion

Emulsions stabilised with decyldimethylphosphine oxide

1) Absence of electrolyte

The effect of decyldimethyl-phosphine oxide (DDPo) concentration on emulsion stability was investigated and the results are illustrated in fig. 1 where log (relative mean droplet volume) is plotted against time for five surfactant concentrations. The curves for light liquid paraffin emulsions are linear suggesting that emulsion breakdown is first-order. This is consistent with the idea that it is the rupture of thin aqueous lamellae separating aggregated oil droplets that is the rate-determining step in the overall process of coalescence. First-order rate constants (k_1) calculated from these plots are given in table 1 and the effect of surfactant concentration on k_1 is illustrated in fig. 2. It can be seen that emulsion stability decreased rapidly below a concentration of ca. 4×10^{-3} mol dm^{-3} which is close to the CMC of DDPo.

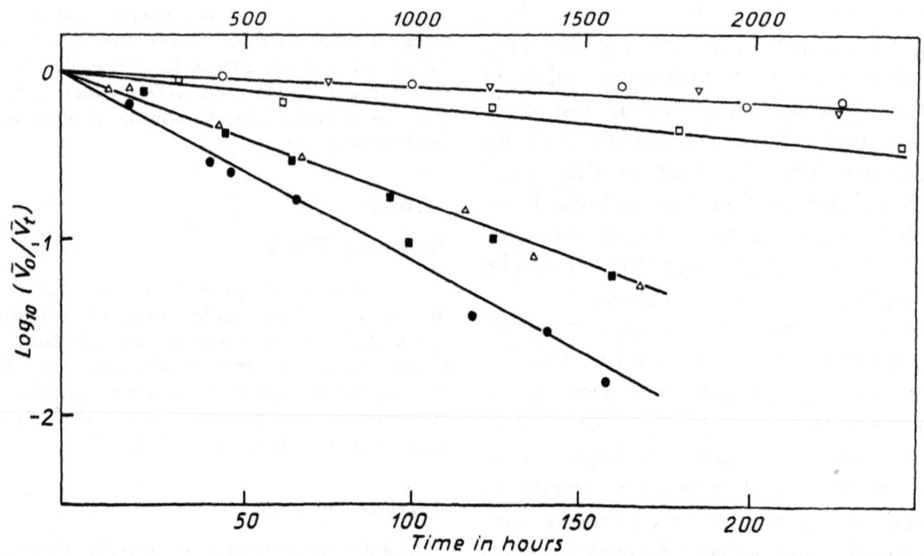

Fig. 1. Variation of mean droplet volume with time for light liquid paraffin/water emulsions stabilised with DDPo. Open symbols refer to upper time scale and close symbols to the lower scale. ○ – 6.03×10^{-3} mol dm^{-3} DDPo; ▽ – 6.03×10^{-3} mol dm^{-3} + 0.5 M KSCN; □ – 3.62×10^{-3} mol dm^{-3}; △ – 3.01×10^{-3} mol dm^{-3}; ■ – 2.67×10^{-3} mol dm^{-3}; ● – 2.26×10^{-3} mol dm^{-3}.

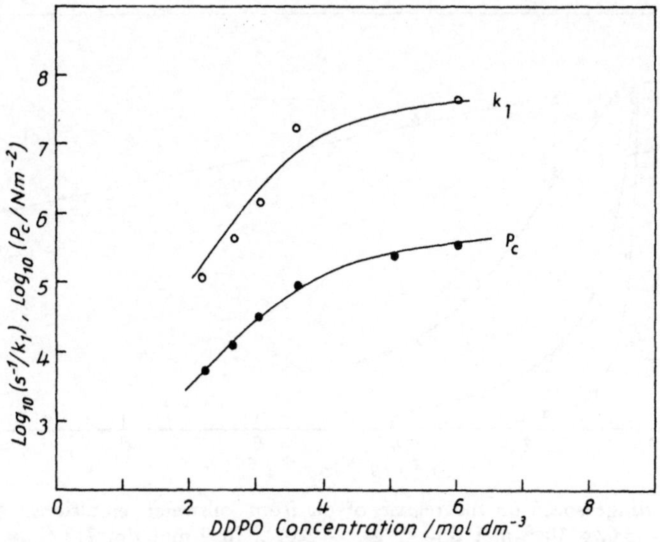

Fig. 2. Effect of surfactant concentration on the first-order rate constant for coalescence and the critical coalescence pressure

The effect of surfactant concentration on stability was also examined by studying accelerated coalescence in a centrifuge. The method of *Smith* and *Mitchell* (7) was adopted. *Smith* and *Mitchell*, using emulsions stabilised with sodium dodecyl sulphate (SDDS), found that the ordered cream or biliquid foam formed at low centrifuge speeds was transparent showing the droplet spacing to be very small (~1nm). DDPo stabilised emulsions had a creamed layer which was always slightly turbid which probably shows that the droplet spacing was larger than for SDDS emulsions. This is consistent with work on Newton black foam films (8) which has shown that films stabilised with DDPo are 1–1.5 nm thicker than those stabilised by SDDS.

Table 1. The effect of DDPo on the coalescence rate constant (k_1) and the critical coalescence pressure (Pc) for oil/water emulsion

DDPo concentration $\times 10^3$/mol dm^{-3}	k_1/sec^{-1}	$t_{\frac{1}{2}}$/days	Pc/Nm^{-2}
6.03	2.3×10^{-8}	348	3.8×10^5
5.10	—	—	2.5×10^5
3.62	5.4×10^{-8}	151	9.9×10^4
3.02	4.6×10^{-7}	17.5	3.1×10^4
2.67	2.25×10^{-6}	3.5	1.2×10^4
2.26	7.2×10^{-6}	1.1	5.4×10^3

For any given speed, the fraction of oil released from DDPo stabilised emulsions was found to be independent of the time of centrifugation after an initial period of about ten minutes had passed. Typical results of the centrifuge experiments are shown in fig. 3 where we have plotted the fraction of free oil produced after centrifugation for thirty minutes against the reciprocal square angular velocity of the rotor. According to *Smith* and *Mitchell* the fraction of free oil is related to the centrifuge speed by

$$R = 1 - [Pc\, A\, \varrho/(\varrho - \varrho_0)\, qm]\omega^{-2}$$

where A is the cross-sectional area of the centrifuge tube; ϱ is the density of the oil; ϱ_0 the density of water; q is the radius of rotation; ω is the angular velocity of the centrifuge; m is the total mass of emulsion particles and Pc is the critical coalescence pressure, that is, the minimum pressure required to cause spontaneous coalescence. A comparison of the limiting ($R \to 1$) slopes of the curves in fig. 3 gives a measure of the relative stability of the emulsions to forced coalescence.

Again emulsion stability is found to decrease rapidly as the stabilised concentration falls below its CMC. Coalescence pressure values calculated from centrifugation curves are given in table 1 and the effect of concen-

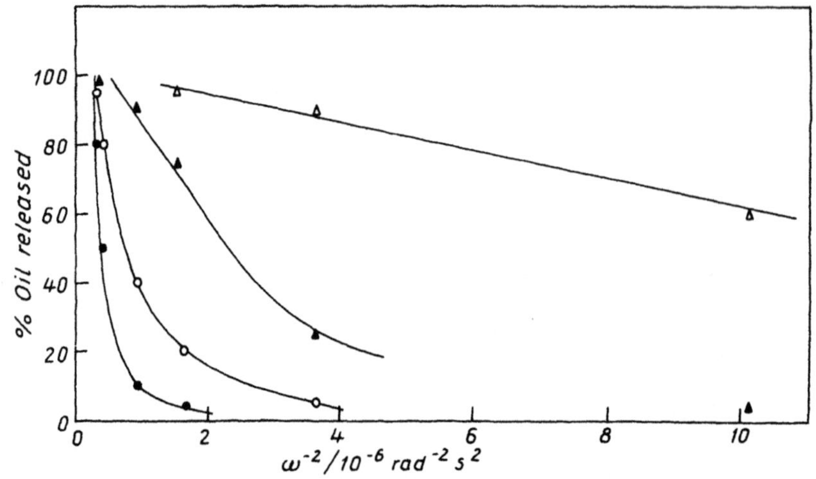

Fig. 3. Effect of centrifuge speed on the release of oil from oil/water emulsions. ● – 6.03×10^{-3} mol dm^{-3} DDPo; ○ – 3.62×10^{-3} mol dm^{-3}; ▲ – 3.01×10^{-3} mol dm^{-3}; △ – 2.26×10^{-3} mol dm^{-3}

tration on Pc is illustrated in fig. 2. It can be seen that both log $(1/k_1)$ and log (Pc) have a very similar dependence upon stabiliser concentration.

The experiments described above illustrate quantitatively the well-known instability of emulsions and thin liquid films observed when the surfactant concentration is reduced below the CMC. Two factors are thought to be important in preventing the rupture of the film separating emulsion droplets, firstly the presence of a complete monolayer which prevents gross or local thinning of the film and secondly a high resistance to dilation which prevents disruption of the monolayer by stretching or deformation of the interfacial film. Thus the rapid onset instability below the CMC might arise from a reduced surface excess concentration or from a reduced interfacial dilational modulus. In principle the surface excess concentration of surfactant and the interfacial dilational modulus can be derived from interfacial tension data.

Fig. 4 shows the effect of DDPo concentration on the interfacial tension of the oil/water interface. The slope $(d\gamma/d \log_{10} C)$ is linear below the CMC showing that there is no measurable change in surface excess con-

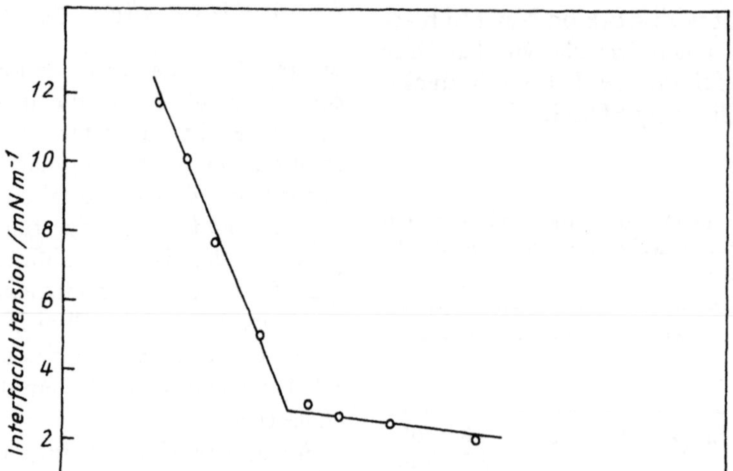

Fig. 4. Effect of DDPo on interfacial tension at the light liquid paraffin/water interface

centration over the concentration region used for emulsion stability studies.

Complete information on the interfacial dilational modulus $\varepsilon(\omega)$ can be obtained by measuring the variation in interfacial tension of an interface subjected to small sinusoidal dilations (9, 10). Unfortunately this type of experiment becomes impracticable when the equilibrium tension is low as is the case with the DDPo/liquid paraffin system. It is however possible to estimate the equilibrium dilational modulus ε_0 from equilibrium interfacial tension data; in any case this is probably the quantity of importance in emulsion stability since current experimental evidence indicated that $\varepsilon(\omega)$ approaches its limiting value at quite low frequencies of deformation (~1 Hz) (9). It is likely that the frequencies of Brownian fluctuations causing coalescence are much higher. ε_0 is defined as[1]).

$$\varepsilon_0 = \frac{d\gamma}{dlnA} = -\frac{d\pi}{dlnA}$$

and can be calculated by differentiating a curve of γ —vs concentration twice. This procedure is however rather inaccurate and *Lucassen* (9) has suggested that better procedure is to calculate ε_0 by adopting a suitable equation of state and using experimental values of the surface pressure π. For example, taking the ideal gaseous equation of state $\pi A = kT$, we find $\varepsilon_0 = \pi$. It is usually found that adsorbed non-ionic surfactants behave ideally at the oil/water interface and many obey the equation:

$$\pi(A - Ao) = kT.$$

This seems an appropriate choice in the present case and on differentiation leads to

$$\varepsilon_0 = \pi + \frac{Ao}{kT}\pi^2.$$

For DDPo Ao is found to be 0.44 nm². Fig. 5 shows that there is a linear relationship between the logarithms of k_1 and Pc and ε_0[2]).

The mechanism of rupture of metastable black films has been considered by *de Vries* (11) and *Derjaguin* and *Gutop* (12). *de Vries* assumed that rupture occurred by the formation of a microscopic hole in the film and found the activation energy for this process to be

$$\Delta Ea = \beta H^2(\gamma + \varepsilon_0)$$

where γ is the film tension, H the film thick-

[1]) From this definition it is apparent that ε_0 is numerical equal to half the Gibbs' elasticity of a thin film (13) and thus for, a thorough it is ignored in the above discussion, ε_0 for an adsorbed layer in a thin film or lamella is dependent upon the film thickness.

[2]) It should be noted for a single interface ε_0 is reduced by micellar relaxation at concentrations above the CMC (9). The present discussion assumes that this is not the case for very thin aqueous films.

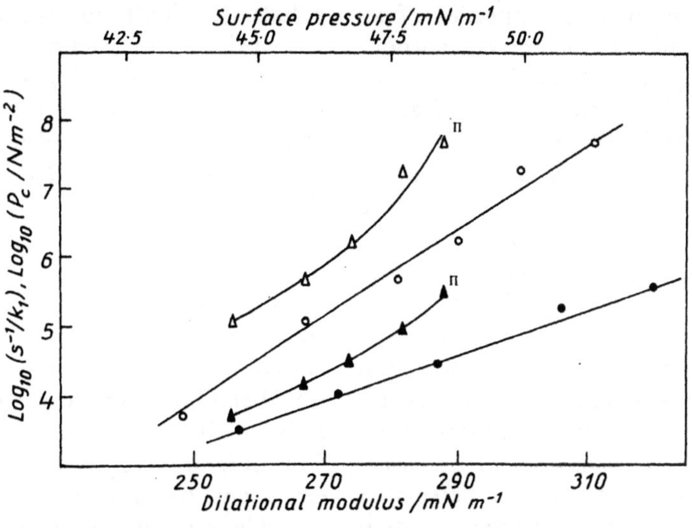

Fig. 5. Variation of coalescence rate constant (open circles) and critical coalescence pressure (closed circles) with interfacial dilational modulus and interfacial pressure

ness at equilibrium and β a geometric constant. de Vries argument involves a number of simplifying assumptions, for example he assumes that γ is independent of the thickness and geometry of the film during the nucleation process. Neverless the conclusion that ΔEa and $(\gamma + \varepsilon_0)$ seems valid since localised thinning must involve stretching or deformation of the film.

In the present case where we are concerned with rupture of thin aqueous lamellae trapped between oil droplets, γ can be ignored since $\varepsilon_0 \gg \gamma$ and so

$$\Delta E_A \propto \varepsilon_0$$

which, on applying the activation energy approach further, leads to

$$\log k_1 \propto \Delta Ea/kT \propto \varepsilon_0.$$

This is in good agreement with the experimental findings.

The argument outlined above while relying on a notional value for ε_0 and on the applicability of the Arrhenius equation to the process of film rupture does illustrate the importance of the interfacial elasticity in emulsion stability. This no doubt arises from its role in maintaining a coherent 'steric' barrier to coalescence.

So far we have only examined the resistance of the interface to dilation and have not considered the role of the shear properties of the interface in stability. These deserve consideration on two counts, firstly because a description of the mechanical properties of an interface or thin film is incomplete without them and secondly because relative motion of neighbouring emulsion droplets may lead to shear stresses at the interface.

Preliminary measurements of the surface shear modulus at the air/water interface have been made. The in-phase component of the surface shear modulus was found to have no measurable value below the CMC, however at concentrations above the CMC rigidity was observed. It is possible that this arose from van der Waals cohesion between the surfactant alkyl chains and thus might not appear at the oil/water interface. This has yet to be ascertained. Nevertheless there is no detectable rigidity at stabiliser concentrations where emulsion stability is appreciable. It is therefore probably that interfacial shear resistance does not play an important part in stability in the present case.

2) *Effect of Electrolytes*

It has been shown elsewhere that oil droplets suspended in DDPo solutions containing sodium chloride acquire a negative charge by adsorption of chloride ions. In this respect DDPo behaves similarly to the related surfactants decylmethylsulphoxide and dodecyldimethylamineoxide (14, 15, 16).

The effect of salt concentration on the zeta-potential of oil droplets suspended in 5×10^{-3} mol dm^{-3} DDPo solutions is shown in fig. 6. For potassium bromide, potassium iodide and potassium thiocyanate the zeta-potential rises slowly with salt concentration and follows the lyotropic series $SCN^- > I^- > Br^-$. With sodium chloride the zeta-potential decreases with increasing salt concentration indicating that the adsorption increases less rapidly with concentration than it does for the other salts. This behaviour is anomalous in the sense that sodium chloride only fits the lyotropic series at high salt concentrations, the results are however similar to those for decylmethylsulphoxide which have been discussed in detail elsewhere (15).

The zeta-potential values shown in fig. 6 are high enough to lead to significant electrical double layer repulsion in some cases and it has been found (8, 15) that stable common-black foam films can be formed from DDPo/salt solutions. This is illustrated in fig. 7 where equilibrium film thickness values are plotted for solutions containing sodium chloride and potassium thiocyanate. In both cases sufficient double layer repulsion is developed to resist the formation of Newton-black films by thinning or disproportion action, these only being formed at electrolyte concentrations > 0.5 mol dm^{-3}.

In comparing the thin film formed between two emulsion droplets with a foam film it should be noted that the van der Waals attraction between droplets is an order of magnitude smaller than that across a foam film. Thus it is to be expected that DDPo stabilised emulsions containing, for example, sodium chloride, should possess an electrical barrier to coagulation which should enhance stability.

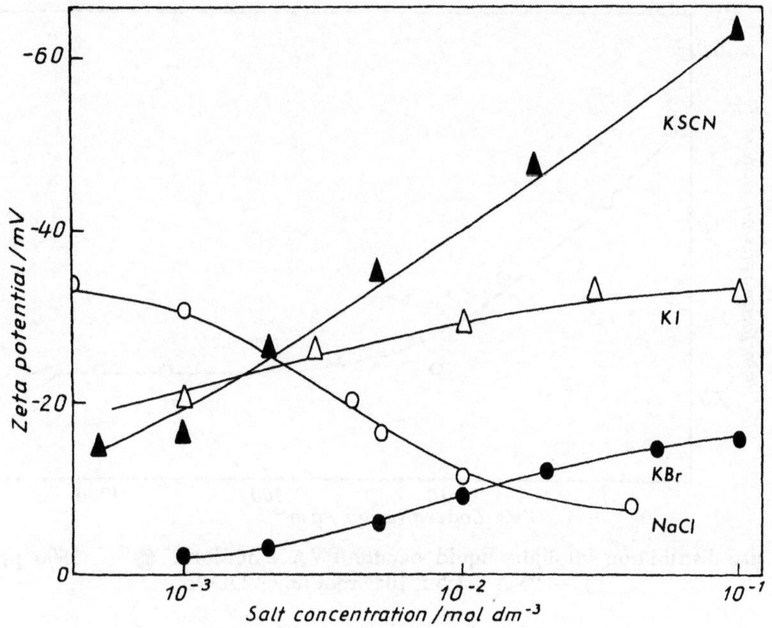

Fig. 6. Effect of salt on zeta-potential for oil droplets dispersed in 5×10^{-3} mol dm^{-3} DDPo solution

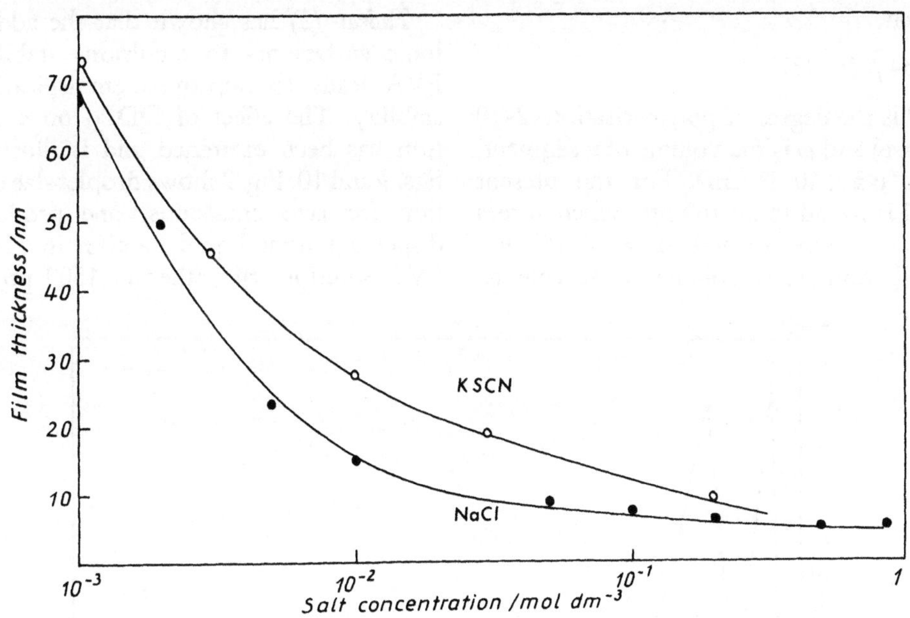

Fig. 7. Effect of salt on the equilibrium thickness of foam films drawn from 5×10^{-3} mol dm^{-3} DDPo solutions

Emulsions Stabilised with Poly(vinylalcohol)

The effect of poly(vinylalcohol) (PVA) on the interfacial tension of the oil-water interface is shown in fig. 8 (measurements were made after the interface had aged for ten minutes). The interfacial tension decreases linearly with concentration until a limiting value of 21 mN m^{-1} is reached. The shape of the curve is similar to that observed for other PVA samples (3, 17). According to *Katchalski* and *Miller* (18) the slope of the

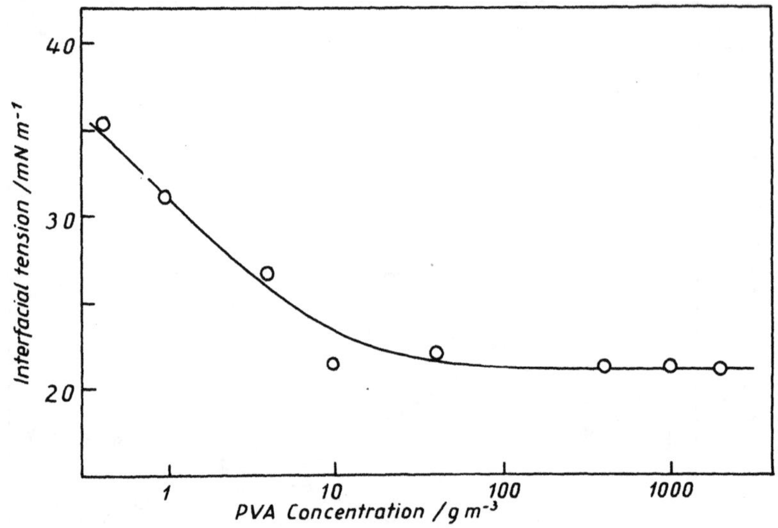

Fig. 8. Droplet size distribution for light liquid paraffin/PVA emulsions. ● – 1500 ppm PVA alone; ○ – PVA + 1.5×10^{-3} mol dm^{-3} DDPo

descending portion of the curve (β) is related to the absorbed layer thickness (δ) by the expression

$$\delta = 0.434\ \beta\ P\ \omega / kT$$

where P is the degree of polymerisation (2410 in this case) and ω is the volume of a segment, taken as 6.3×10^{-23} cm^3. For the present sample δ is found to be 160 nm which agrees well with an experimental value of 152 nm calculated from cream volume measurements.

Poly(vinylalcohol) + DDPo

Tadros (3) has shown that the addition of ionic surfactants to emulsions stabilised by PVA leads to improved emulsification and stability. The effect of DDPo on emulsification has been examined and is illustrated in figs. 9 and 10. Fig. 9 shows droplet size distribution for two emulsions, one produced by dispersing light liquid paraffin in 1500 ppm PVA solution, the other in 1500 ppm PVA

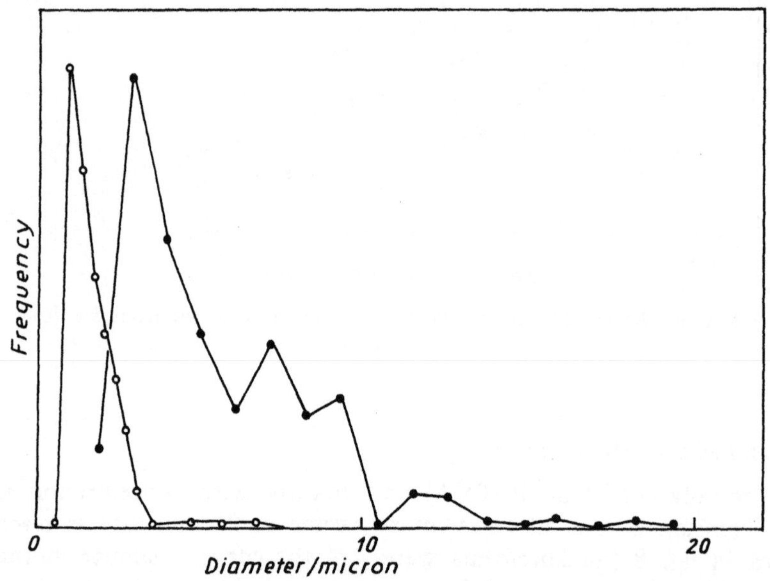

Fig. 9. Effect of DDPo on droplet size for light liquid paraffin/PVA emulsions. Solid lines – quartile diameters; broken line – mean surface area diameter. ● – 6×10^{-3} mol dm^{-3} DDPo alone.

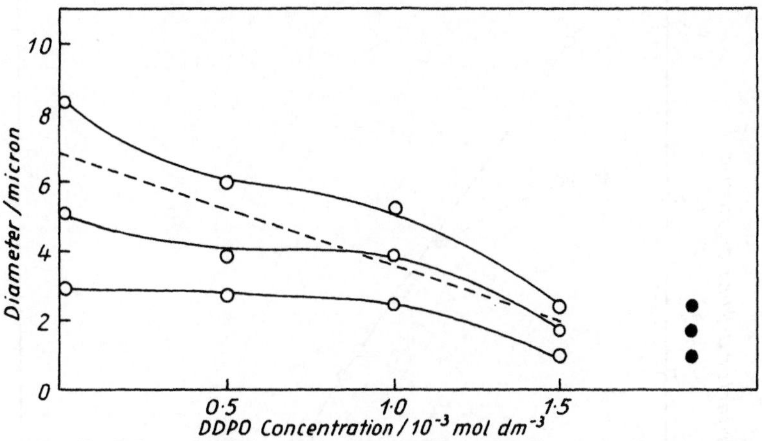

Fig. 10. Effect of PVA on the interfacial tension at the oil/water interface

containing 1.5×10^{-3} mol dm^{-3} DDPo. The added surfactant has two effects, it reduces the modal diameter of the emulsion and greatly reduces the polydispersity. The effect of intermediate surfactant concentrations is illustrated in fig. 10 where the quartile diameters and the mean surface area diameter are shown. The mean surface-area diameter decreases linearly with concentration, the behaviour of the quartiles however is more interesting. At low surfactant concentrations the upper quartile diameter is reduced while the median and lower quartile are unaffected. At higher concentrations all three parameters are reduced. This probably indicates that low concentrations of DDPo have no effect on the diameter of primary droplets but simply prevents their recombination. At higher concentrations it appears that adsorption of DDPo is rapid enough to reduce the dynamic interfacial tension during emulsification and leads to a reduction in the size of primary droplets. At a concentration of added DDPo of 1.5×10^{-3} mol dm^{-3} the droplet size distribution is similar to that found using 6×10^{-3} mol dm^{-3} DDPo alone. Thus the addition of DDPo to PVA solution leads to improved emulsification and homogenisation.

Tadros (3) found that emulsions prepared using surfactant-PVA mixtures were, in some cases, more stable than those prepared using polymer or surfactant alone and attributed this stability to the presence of polymer surfactant complexes. It is now well established (3, 4, 19, 20) that ionic surfactants associate with PVA to form polyelectrolyte-like complexes which exist in bulk solution and at the air-liquid, liquid-liquid and solid-liquid interface. It is thought that association occurs primarily by hydrophobic bonding between the surfactant alkyl chain and the polymer backbone. It would be expected that a non-ionic surfactant such as DDPo, which differs from simple ionic surfactants only in its unionised headgroups, should behave similarly if hydrophobic bonding is the cause of association. However we find that this is not the case, viscosity measurements show that PVA solutions and PVA solutions containing DDPo have identical intrinsic viscosities which suggests that association does not occur in the bulk.

Tadros (3) found that interfacial tension curves for PVA-surfactant mixtures had a form different to those found for PVA or surfactant alone and attributed this to adsorption of the polyelectrolyte-like complex. Again this type of behaviour was not observed with DDPo-PVA mixtures. Interfacial tension values for DDPo solutions and DDPo solutions containing 1500 ppm PVA are shown in fig. 11. At concentrations higher than ca. 5×10^{-4} mol dm^{-3} the curves are superimposed and show that adsorption and micellisation of DDPo are unaffected by the polymer. At lower DDPo concentrations the interfacial tension levels off at the plateau value found for PVA solutions (ca. 21 mNm^{-1}). It would appear that the surfactant displaces the polymer from the interface. As the present sample of PVA was different to that used by *Tadros* (having a molecular weight of 125 kg mol^{-1} and 20% acetate groups, whereas *Tadros*' sample had a molecular weight of 42 kg mol^{-1}

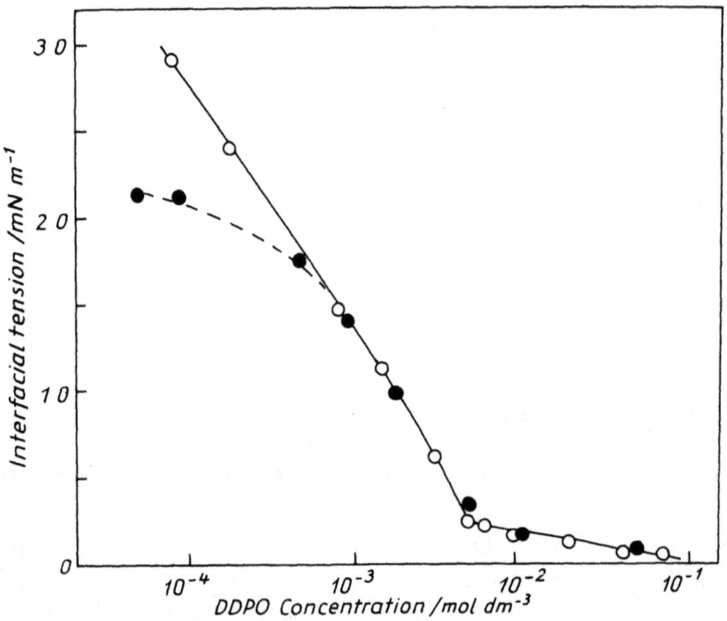

Fig. 11. Effect of PVA on the interfacial tension at the oil/aq. PVA solution interface. ● – 1500 ppm PVA+DDPo; ○ – DDPo alone

and contained 12% acetate groups) it was worthwhile to check that it had formed association complexes with ionic surfactants. This was however found to be the case. Typical interfacial tension data are shown in fig. 12. The curves for cetyltrimethylammonium bromide (CTAB) and sodium alkyl sulphates are similar to those obtained for CTAB and sodium dodecylbenzenesulphonate by *Tadros*. It is also worth noting that it is not the relatively short alkyl chain length of DDPo which inhibits association since sodium decylsulphate behaves similarly to the longer chain ionic surfactants.

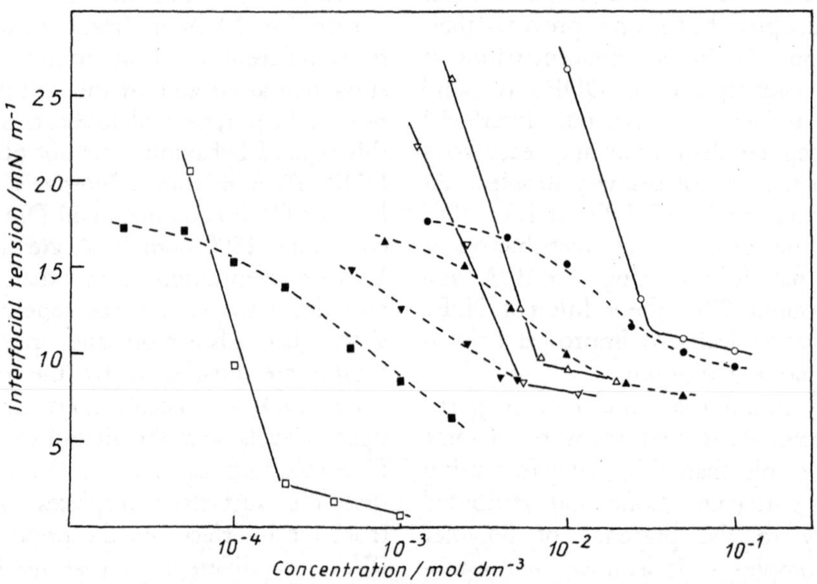

Fig. 12. Effect of ionic surfactant on the interfacial tension at the oil/water interface. Open symbols – surfactant alone; closed symbols surfactant +1500 ppm PVA. ○ – sodium decyl sulphate; △ – sodium dodecyl sulphate; ▽ – sodium decyl sulphate +0.2 mol dm⁻³ NaCl. □ – CTAB

The results described above cast doubt upon the idea that PVA-surfactant association arises through hydrophobic bonding and suggest that the surfactant polar group plays a part. A similar conclusion has been reached by *Schwuger* (21) who found that polymers such as poly(vinylpyrrolidone), polyglycolethers and PVA formed strong complexes with anionic surfactants but weaker complexes with cationics. In the case of polyethylene glycol he found that the tendency to form complexes increased with the number of glycol units. Both of these findings point to the conclusion that polar regions of the polymer and surfactant molecules are involved in association.

The interfacial tension data show that DDPo displaces the polymer from the oil/water interface. In view of this we have attempted to assess the effect of added DDPo on the stability of PVA stabilised emulsions using the centrifuge method. Unfortunately the release of oil from PVA stabilised emulsions was found to be time-dependent over practicable centrifugation periods and so the analysis of *Smith* and *Mitchell* could not be applied to the results. Thus we have chosen to compare the amount of oil released after thirty minutes centrifugation at 16,000 r.p.m. The results are given in table 2 together with an estimate of a critical coalescence pressure (Pc^*) calculated after the same period. It can be seen that emulsions containing added DDPo are less stable than those containing PVA alone and, in the case of higher DDPo concentrations, less stable than those containing DDPo alone. This is the opposite of the effect

Table 2. The effect of DDPo on the centrifugal stability of oil/PVA solution emulsions

Stabiliser	Percent free oil at 16,000 R.P.M.	Pc^*/Nm^{-2}
1500 ppm PVA	5%	$>10^6$
$+2.5\times10^{-4}$ mol dm^{-3} DDPo	5%	$>10^6$
$+5.0\times10^{-4}$ mol dm^{-3} DDPo	50%	$\sim2.5\times10^5$
$+1.5\times10^{-3}$ mol dm^{-3} DDPo	70%	$\sim1.5\times10^5$
$+4.0\times10^{-3}$ mol dm^{-3} DDPo	90%	$\sim10^5$
$+6.0\times10^{-3}$ mol dm^{-3} DDPo	90%	$\sim10^5$
6.0×10^{-3} mol dm^{-3} DDPo alone		3.8×10^5

of ionic surfactants on stability and, from a practical view, demonstrates the importance of choosing an emulsifying additive which is compatible with the polymeric stabiliser used.

Acknowledgements

I would like to thank the Science Research Council for their financial support; Professor *R. H. Ottewill* for his gift of a sample of decyldimethylphosphine oxide and Dr. *S. S. Davis* for his encouragement of this investigation.

Summary

The stability and properties of oil-in-water emulsions stabilised by a simple non-ionic surfactant, decyldimethylphosphine oxide (DDPo) and by mixtures of DDPo with poly(vinylalcohol) and electrolytes have been investigated.

For emulsion stabilised by surfactant alone the results confirm the idea that it is the high interfacial elasticity arising from adsorbed surfactant that determines stability. The variation of stability with surfactant concentration appeared to arise from the accompanying variation of the interfacial dilational modulus.

Unlike simple ionic surfactants it was found that DDPo did not associate with poly(vinylalcohol) and although its addition to pVA solutions greatly improved emulsification, emulsion stability was impaired. Interfacial tension data suggested that the surfactant displaced the polymer from the interface.

Zusammenfassung

Es wurden Stabilität und Eigenschaften von Öl-in-Wasser-Emulsionen untersucht, die entweder durch das nichtionogene Tensid Decyldimethylphosphinoxid (DDPo) oder durch Mischungen von DDPo mit Polyvinylalkohol und Elektrolyten stabilisiert waren.

Für Emulsionen, die mit dem Tensid allein stabilisiert waren, bestätigen die Ergebnisse die Annahme, daß die Stabilität durch die hohe Grenzflächenelastizität des adsorbierten Tensids bestimmt wird. Die Stabilitätsänderung mit der Tensidkonzentration scheint mit Änderungen des Dilatationsmodulus in der Grenzfläche zusammenzuhängen.

Im Gegensatz zu einfachen ionogenen Tensiden erfolgt zwischen DDPo und dem Polyvinylalkohol keine Assoziation, und obwohl der Zusatz von DDPo zur Polyvinylalkohollösung die Emulsifizierung wesentlich verbesserte, blieb die Emulsionsstabilität unverändert. Die Daten über die Grenzflächenspannung legen die Annahme nahe, daß das Tensid Polymermoleküle von der Grenzfläche verdrängt.

References

1) *Florence, A. T.* and *J. A. Rogers*, J. Pharm. Pharmac., **23**, 153 (1971).

2) *Clunie, J. S., J. F. Goodman,* and *B. T. Ingram*, in Surface and Colloid Science, vol. III, Ed. *E. Matijévic*, (New York 1971).

3) *Tadros, Th. F.*, S.C.I. Symposium Brunel University 1974 Theory and Practice of Emulsion Technology.

4) *Tadros, Th. F.*, J. Colloid Interface Sci. **46**, 3, 528 (1974).

5) e.g. *Adamson, A. W.*, Physical Chemistry of Surfaces 2nd Ed. (London 1967).

6) *Sherriff, M.* and *B. Warburton*, Polymer **15**, 253 (1974).

7) *Smith, A. L.* and *D. P. Mitchell*, S.C.I. Symposium Brunel University 1974 Theory and Practice of Emulsion Technology.

8) *Buscall, R.*, Thesis, Bristol University (1973).

9) *Lucassen, J.* and *D. Giles*, J.C.S. Faraday I, **71**, 217, (1975).

10) *Lucassen, J.*, Faraday Disc. Bristol 1975 Physical Adsorption in Condensed Phases.

11) *de Vries, A. J.*, Rec. Trav. Chim., **77**, 383 (1958).

12) *Derjaguin, B. V.* and *Yu. V. Gutop*, Colloid J. USSR, **24**, 370 (1962).

13) *Lucassen, J.* and *E. G. Lucassen-Reynders*, J. Colloid Interface Sci., **25**, 496 (1967).

14) *Buscall, R., R. B. Donaldson, R. H. Ottewill*, and *D. Segal*, S.C.I. Symposium on Foams. Brunel University, 1975.

15) *Buscall, R.* and *R. H. Ottewill*, Adv. Chem., **144**, "Monolayers", Ed. *E. D. Goddard*, 1975 (in press).

16) *Donaldosn, R. B.*, Thesis, Bristol University (1971).

17) *Lankveld, J. M. G.* and *J. Lyklema*, J. Colloid Interface Sci., **41**, 466 (1972).

18) *Katchalsky, A.* and *I. Miller*, J. Phys. Chem. **55**, 1182 (1951).

19) *Jones, M. N.*, J. Colloid Interface Sci., **23**, 36 (1967).

20) *Arai, H., M. Murata*, and *K. Shinoda*, J. Colloid Interface Sci., **37**, 223 (1971).

21) *Schwuger, M. J.*, J. Colloid Interface Sci., **43**, 491 (1973).

Author's address:
Dr. *R. Buscall*
Physical Chemistry Department, School of Chemistry
The University
Bristol BS 8 1 TS (England)

Progr. Colloid & Polymer Sci. **63**, 27–29 (1978)

ISSN 0340-255 X

Department of Chemistry, K. R. College, Mathura and Department of Chemistry, Agra College, Agra (India)

Role of mixed emulsifiers in emulsion stability

A. K. Goswami and *P. Bahadur*

With 3 figures

(Received June 21, 1976)

Introduction

It has long been recognised that mixed emulsifying agents very often yield more stable emulsions than do single ones, and, in most cases, the high stability is accounted for the complex formation at the interface, which results in the interfacial film of greater strength so prevents coalescence of droplets. However, only a few workers (1–4) having attempted the problem, there still remain some aspects to be explored. In this communication we have dealt with the ionic surfactants as emulsifiers and both o/w and w/o type promoting emulsifiers have been selected.

Materials

The detergents used were cetyl trimethyl ammonium bromide (CTAB) (cationic), dioctylester of sodium sulphosuccinic acid (OSS) (anionic) and stannous oleate (anionic). All these were BDH or sigma products and were of high degree of purity. Xylene was AnalaR BDH, water used was always doubly distilled and the inorganic salts used as flocculating agents were all BDH or EM chemicals of AnalaR grade.

The xylene-in-water emulsions were prepared by suspending 5% volume of xylene in water containing 0.01 M KCl. The emulsifiers CTAB and OSS were dissolved in aqueous phase whereas stannous oleate was in the oil phase. The concentration of emulsifiers was always 1 mM. For emulsions with detergent mixtures, the concentration of each detergent was 0.5 mM. Five emulsions were prepared and were characterised as (I) CTAB system (II) OSS system (III) Sn (ii) oleate system (IV) CTAB + OSS system (V) CTAB + Sn (ii) oleate system.

The content for the emulsion was first hand shaken for about fifteen minutes and finally homogenized twice in a stainless steel homogenizer (Schaar Scientific Co., Chicago). The freshly prepared emulsion was stored in stoppered glass tubes and the creamed layer formed was skimmed off. All the emulsions were prepared under identical conditions.

Electrophoretic mobilities of the dispersed emulsion droplets were measured microelectrophoretically using Northrup-Kunitz type cell by diluting the emulsions ten times with 0.01 M KCl. The mobility was calculated by determining the particle velocity under unit potential gradient. The zeta-potential values were calculated directly by the use of Helmholtz-Smoluchowski equation.

All the emulsions resulted almost same particle size equal to 2 microns diam.

Results and discussion

Effect of electrolytes on zeta-potential

Stable oil-in-water type emulsions with negative charge were obtained with OSS. Stannous oleate was rather poor emulsifier for xylene-in-water emulsions, the charge on the globules of such systems was also negative. With (CTAB + OSS), however, the emulsions with very low values of negative zeta potential were prepared and all such systems were having very low stability as examined microscopically. The effect of inorganic salts with varying cations on zeta-potentials of such systems was investigated and the zeta potential – log C curves plotted (figs. 1 and 2). All the electrolytes decreased the zeta potential values and polyvalent ions were found very effective with which the emulsions were brought at isoelectric point. On further addition of the salt, the zeta potential values were increased but the systems now acquire positive charge. Very stable emulsions could be prepared by using CTAB + Sn (ii) oleate as emulsifiers. It is thus concluded that the mixture of anionic and cationic surfactants form

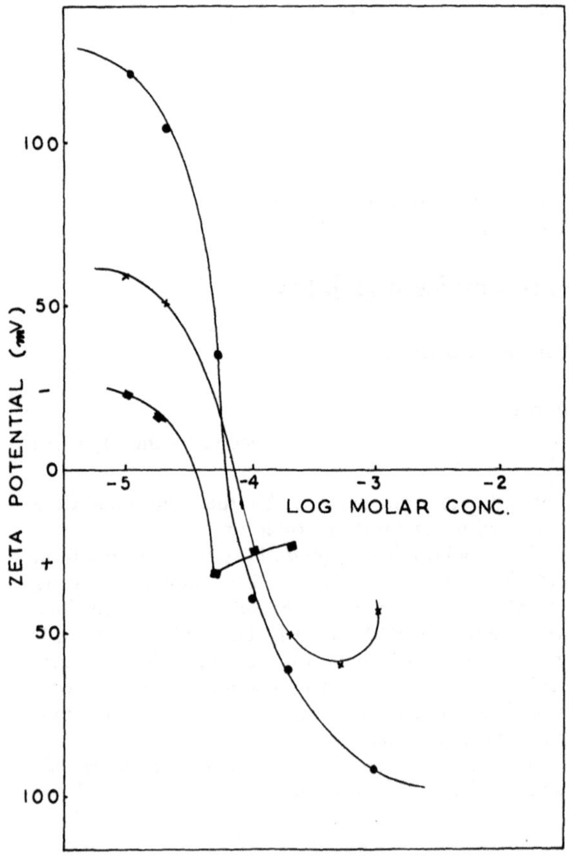

Fig. 1. Zeta potential as a function of log-concentration of $Th(NO_3)_4$ for negatively charged emulsions ■ CTAB + OSS, × Sn(II) oleate, ● OSS

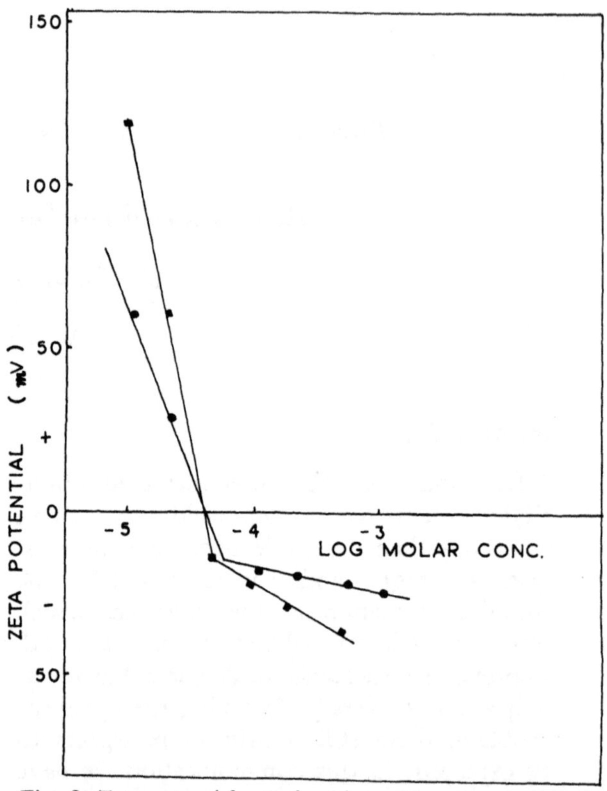

Fig. 2. Zeta potential as a function of log-concentration of $K_4Fe(CN)_6$ for positively charged emulsion ■ Sn(II) oleate + CTAB, ● CTAB

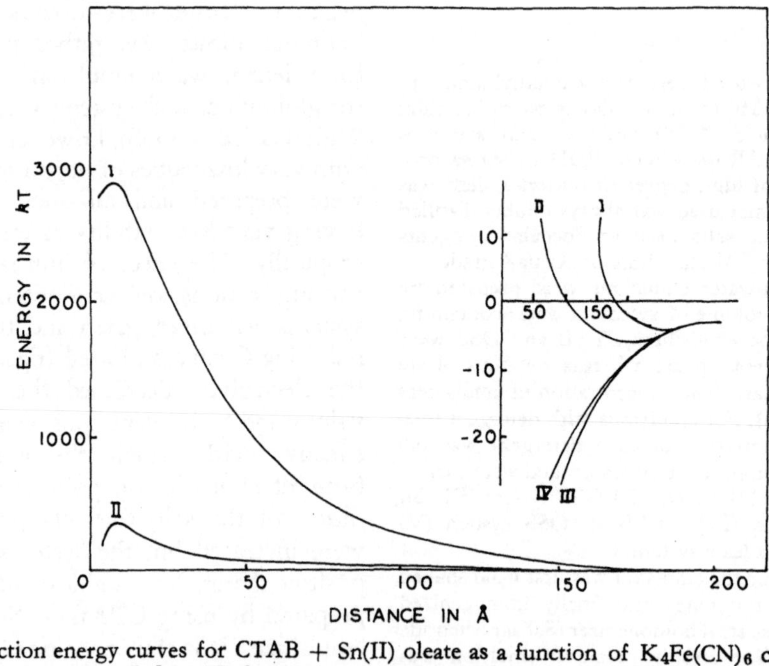

Fig. 3. Interaction energy curves for CTAB + Sn(II) oleate as a function of $K_4Fe(CN)_6$ concentration. I. 1×10^{-5} M, II. 2×10^{-5} M, III. 5×10^{-5} M IV. 1×10^{-4} M

stable emulsions when both these are present in separate phases. The presence of both the systems in a single phase form detergent complex before emulsification and it has got poor adsorption characteristic and thus unstable emulsions are formed.

Interaction energy and stability of emulsions

In view to test the validity of DLVO theory to the present systems, the interaction energies have been calculated as a function of interparticle distance. All the studies were made almost at the same ionic strength. High energy hump could be seen in energy diagrams for the cases where the values of zeta potential were greater than 30 mV. It is rather improbable for the flocculation of the systems by crossing such high energy barriers when the particles are sufficiently close together. We demonstrate here the flocculation in the secondary minima when the particles are sufficiently apart such type of flocculation will be by nature reversible and the flocculated droplets may be further redispersed by some mechanical means. For the systems where the zeta potentials were less than 30 mV, attraction predominates over repulsion at all interparticle distances and here one may find the flocculation even in the primary minima. A representative set of energy curves (fig. 3) explains both types of flocculation.

Acknowledgments

We are deeply grateful for the financial support of the University Grant Commission, India, which made this work possible. The authors wish to thank Dr. *T. C. Sharma*, Principal, and Dr. *R. C. Saraswat*, Head, Chemistry Department, K. R. College, Mathura, India, for many fruitful discussions and encouragement.

Summary

The stability of emulsions stabilized by ionic detergents and their binary mixtures was studied and the relative emulsifying powers of detergents determined. The validity of DLVO theory to the emulsion systems was tested. Emulsions with zeta potentials less than 30 mV were flocculated rapidly. For emulsions with high zeta values the flocculation could be seen in secondary minima. Emulsion with detergent mixtures (anionic + cationic) were less stable (than emulsions prepared by the use of single detergent as emulsifier), if both the surfactants were initially present in the same phase. Stable emulsions with detergent mixtures were obtained by dissolving them in different phases.

Zusammenfassung

Es wurde die Stabilität von Emulsionen untersucht, welche durch ionische Tenside und Mischungen derselben stabilisiert waren. Gleichzeitig wurde das Emulgiervermögen der Tenside untersucht und die Gültigkeit der DLVO-Theorie getestet. Emulsionen mit Zetapotentialen unter 30 mV wurden rasch geflockt. Bei hohen Zetapotentialen wurde Flockung im sekundären Minimum beobachtet. Emulsionen mit Mischungen von anionischen und kationischen Tensiden waren weniger stabil als Emulsionen, die nur mit einem einzigen Tensid als Emulgator hergestellt waren, wenn beide Tenside von Anfang an in der gleichen Phase zugegen waren. Stabile Emulsionen mit Tensidmischungen wurden dagegen beobachtet, wenn die einzelnen Tenside in den verschiedenen Phasen gelöst wurden.

References

1) *Schulman, J. H., E. G. Cockbain*, Trans. Faraday Soc. **36**, 651 (1940).

2) *Dickinson, W., Iball*, J. Research (London) **1**, 614 (1948).

3) *Cockbain, E. G., T. S. Mc Roberts*, J. Colloid Sci. **8**, 440 (1956).

4) *Sastry, T. G.*, Ph. D. Thesis, Agra Univ. (1969).

Authors' addresses:

Ashok K. Goswami
Department of Chemistry,
K. R. College
Mathura 281001
India

P. Bahadur
Chemistry Department,
Agra College,
Agra 282002

Progr. Colloid & Polymer Sci. **63**, 30–32 (1978)

ISSN 0340-255 X

Chemistry Department, K. R. College, Mathura (India) and Chemistry Department, Agra College, Agra (India)

Studies on interaction of cationic surfactants with cholesterol coated oil droplets dispersed in water

P. M. Gupta, P. Bahadur, and *S. N. Srivastava*

With 2 figures and 2 tables

(Received July 7, 1977)

Introduction

Interfacial properties of bilesalts and sterols are biologically important and have got some role in life process. *Tien* and *Diana* (1) have described the formation of bilayers from cholesterol dissolved in hydrocarbon in contact with an aqueous solution of cetyl trimethyl ammonium bromide (CTAB). *Sweeney* and *Blank* (2) have studied the electrical properties of the film formed by cholesterol and CTAB. Extensive studies on mixed monolayers of cholesterol and CTAB have been made by *Ahmad* (3).

In our previous communication (4) we have studied the interfacial properties of sodium deoxycholate and its interaction with some cationic surfactants.

Experimental

Cholesterol as emulsifier was a BDH product. The cationic detergents viz., lauryl pyridinium chloride (LPC), cetyl trimethyl ammonium bromide (CTAB), cetyl pyridinium bromide (CPB) and cetyl dimethyl benzyl ammonium chloride (CDBAC), were all BDH chemicals and used without further purification. AnalaR Toluene was used as the oil phase and double distilled water was used throughout the measurements,

The o/w emulsions were prepared by shaking the oil-water mixture with a ratio o/w of 1:19. The emulsifier concentration was 0.5% and the ionic strength 0.01 maintained by KCl. The mixture was passed twice through a stainless steel homogenizer which gave emulsion droplets of 2 μm average size.

Results

The zeta potential of emulsion droplets and stability

The average mobility of the droplets in cholesterol stabilized emulsion was about $3.8 \cdot 10^4$ μ/sec/volt/cm which led to a zeta potential of 59.75 mV.

The effect of cationic detergents on the zeta potential of the system is shown in figure 1. In general all the detergents reduced the zeta potential significantly. The flocculating concentrations are $8.41 \cdot 10^{-7}$ M, $5.96 \cdot 10^{-7}$ M, $5.62 \cdot 10^{-7}$ M and $4.46 \cdot 10^{-7}$ M for LPC, CTAB, CPB and CDBAC respectively.

Potential energy curves were constructed using approximations in calculating the potential energy of

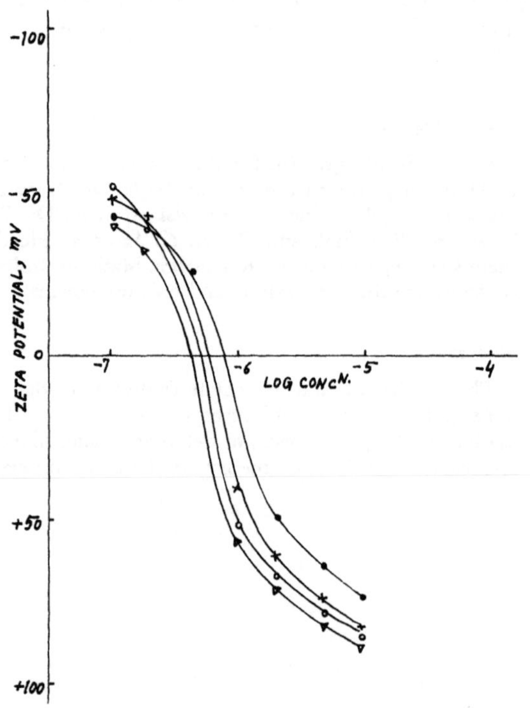

Fig. 1. Zeta potential (ζ) - log C curves for ● LPC, × CTAB, ○ CPB, △ CDBAC

repulsion (V_R) and attraction (V_A) for interacting droplets with same size (fig. 2). The data are reported in table 1. The value of V_R can be calculated with sufficient accuracy from the equation (5) ($\varkappa a \gg 1$):

$$V_R = \frac{ta\psi_0^2}{2} \ln\left[1 + \exp(-\varkappa H)\right] \quad [1]$$

where H is the distance between two adjacent droplets. The potential (ψ_0) of the diffuse double layer is assumed identical with zeta potential.

Schenkel and *Kitchener*'s (6) modified form of the Hamaker equation for V_A, where droplets are in close proximity, i.e. $H_0 \ll a$, was used and no correction was applied for the layer of adsorbed emulsifier. For distances greater than 0.1 λ, ($\lambda \approx 10^{-5}$ cm, wave length of intrinsic electronic oscillations of the atoms) the retardation effect reduces the magnitude of the V_A. Thus

$$V_A \approx \frac{-Aa}{12\,H}\left[\frac{\lambda}{\lambda + 3{\cdot}54\,\pi H}\right] \quad \text{for } H < 150\ \text{Å}$$

$$V_A \approx \frac{-Aa}{\pi}\left[\frac{2{\cdot}45\,\lambda}{120\,H^2} - \frac{\lambda^2}{1045\,H^3} + \frac{\lambda^3}{5{\cdot}62 \times 10^4\,H^4}\right] \quad \text{for } H > 150\ \text{Å} \quad [2]$$

where A is the Hamaker constant when the adsorbed emulsifier layer has no significant thickness, the value of $2.8\cdot10^{-13}$ erg for pure toluene droplets dispersed in water was taken for the Hamaker constant.

From the data in table 1 and figure 2 some conclusions can be drawn: The energy maxima decrease generally with increasing concentration of detergent up to charge reversal (the concentrations of surfactants used to reverse the charge) and increase with further addition of detergents. The net potential energies are always greater than 25 kT which are sufficiently high for stable emulsions.

Despite the high energy barriers the emulsion was, however, flocculated at sufficient rate as examined microscopically. Therefore, one can assume that the flocculation is encountered by the secondary minima. Sufficiently deep secondary minima were calculated in all cases (table 1). The reversible nature of flocculation was revealed by redispersing flocculated emulsions by mechanical means. Creaming or sedimentation may be the secondary factors in flocculating the systems.

Table 1. Energy data for cholesterol stabilized Toluene in water emulsion flocculated by cationic detergents

Detergent conc. in mole/lit.	Zeta potential (mV)	Height of maxima in kT	Distance at $V=0$ (Å)	Depth of secondary minima in kT
LPC				
$1.0\cdot10^{-7}$	39.9	464	130	11.5
$2.0\cdot10^{-7}$	38.4	412	125	12.0
$5.0\cdot10^{-7}$	25.5	52	100	15.5
Charge reversal				
$2.0\cdot10^{-6}$	50.1	935	150	11.0
$5.0\cdot10^{-6}$	63.0	1773	160	10.5
$1.0\cdot10^{-5}$	74.5	2678	175	10.0
CTAB				
$1.0\cdot10^{-7}$	47.0	755	140	11.0
$2.0\cdot10^{-7}$	42.0	741	130	11.5
Charge reversal				
$1.0\cdot10^{-6}$	42.6	564	130	11.5
$2.0\cdot10^{-6}$	60.0	1560	160	10.5
$5.0\cdot10^{-6}$	72.0	2471	170	10.0
$1.0\cdot10^{-5}$	82.0	3351	185	9.5
CPB				
$1.0\cdot10^{-7}$	52.6	1081	155	10.5
$2.0\cdot10^{-7}$	38.6	409	110	14.0
Charge reversal				
$1.0\cdot10^{-6}$	52.8	1096	145	11.0
$2.0\cdot10^{-6}$	66.0	2000	175	10.5
$5.0\cdot10^{-6}$	78.5	3031	185	10.0
$1.0\cdot10^{-5}$	87.3	3892	195	9.5
CDBAC				
$1.0\cdot10^{-7}$	37.6	383	130	12.0
$2.0\cdot10^{-7}$	32.4	224	115	15.5
Charge reversal				
$1.0\cdot10^{-6}$	57.0	1359	170	10.5
$2.0\cdot10^{-6}$	70.2	2322	130	10.0
$5.0\cdot10^{-6}$	81.0	2749	185	9.5
$1.0\cdot10^{-5}$	87.0	3855	195	9.5

Binding parameters at iso-electric point

The binding parameters at iso-electric point of the system were calculated following the treatment given by *Wilkins* (7), *Ottewill* (8), and *Srivastava* (9–10). Details of equations used for this purpose have been reported elsewhere (4). The values of adsorption constants, charge density in Stern layer, free energy of adsorption, number of binding sites available etc. are recorded in table 2.

The values of ΔG as recorded in table 2 are of the order of 11–12.5 Kcals/mole. ΔG increases with chain length of the detergents, this implies that charge reversal takes place at

Table 2. Parameters for detergent binding

Detergents	Flocculation concentration (moles/lit.)	$N_1 \times 10^{12}$ groups/cm^2	σS 10^{-3}	$-\Delta G$ K cals/mole
LPC	$8.41 \cdot 10^{-7}$	9.00	4.14	12.23
CTAB	$5.96 \cdot 10^{-7}$	9.43	4.15	11.95
CPB	$5.62 \cdot 10^{-7}$	9.90	4.13	11.69
CDBAC	$4.46 \cdot 10^{-7}$	10.42	4.14	11.63

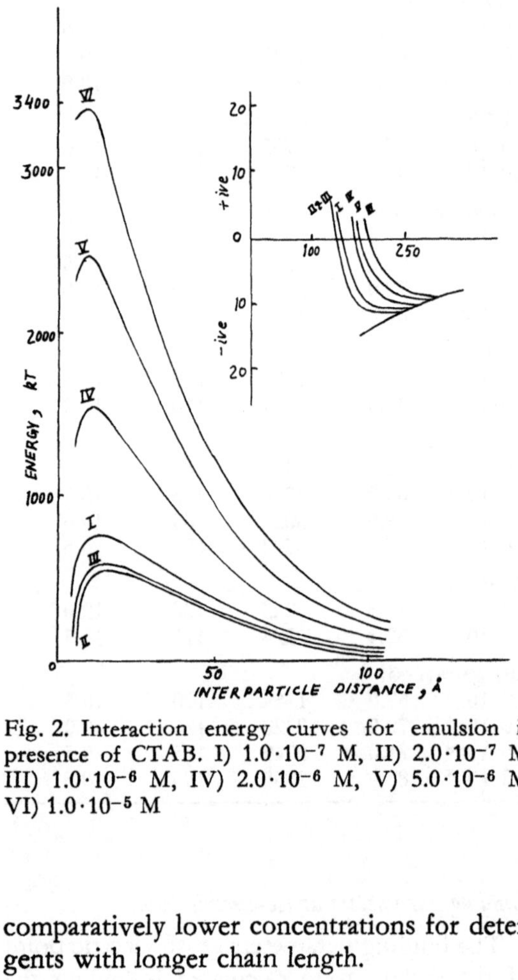

Fig. 2. Interaction energy curves for emulsion in presence of CTAB. I) $1.0 \cdot 10^{-7}$ M, II) $2.0 \cdot 10^{-7}$ M, III) $1.0 \cdot 10^{-6}$ M, IV) $2.0 \cdot 10^{-6}$ M, V) $5.0 \cdot 10^{-6}$ M, VI) $1.0 \cdot 10^{-5}$ M

comparatively lower concentrations for detergents with longer chain length.

Summary

Cholesterol stabilized oil-in-water emulsions were flocculated by cationic surfactants. Very low concentrations of surfactants flocculated the system completely, the flocculating values being $8.41 \cdot 10^{-7}$ M, $5.96 \cdot 10^{-7}$ M, $5.62 \cdot 10^{-7}$ M, and $4.46 \cdot 10^{-7}$ M for LPC, CTAB, CPB and CDBAC respectively. Further addition of surfactant showed marked stability. The results were based on electrophoretic measurements and interpreted in the light of DLVO theory. The efficiency of surfactants in flocculating the systems depends on their chain length.

Zusammenfassung

Cholesterol-stabilisierte Öl/Wasser-Emulsionen wurden durch kationische Tenside geflockt. Bei sehr geringen Konzentrationen wird das System geflockt: 8.41×10^{-7} M für LPC, CTAB, CPB und CDBAC (Bezeichnung s. S. 30). Bei weiterer Tensidzugabe tritt Stabilisierung ein. Die Ergebnisse werden auf Grund von elektrophoretischen Messungen mit der DLVO-Theorie interpretiert. Die flockende Wirkung der Tenside hängt von der Kettenlänge ab.

References

1) *Tien, H. T., A. L. Diana*, J. Colloid & Int. Sci. **24**, 287 (1967).

2) *Sweeney, G. D., M. Blank*, J. Colloid & Int. Sci. **42**, 410 (1973).

3) *Ahmad, J.*, J. Colloid Int. Sci. **49**, **1** (1974).

4) *Gupta, P. M., P. Bahadur, S. N. Srivastava*, J. Colloid & Int. Sci. **59**, 45 (1977).

5) *Derjaguin, B. V., M. Kussakov*, Acta Phys. Chim. **10**, 25, 155 (1939).

6) *Schenkel, J. N., J. A. Kitchener*, Trans. Faraday Soc., **56**, 161 (1960).

7) *Wilkins* et al., J. Theoret. Biol. **2**, 173, 186 (1962).

8) *Ottewill, R. H., M. C. Rastogi, A. Watanabe*, Trans. Faraday Soc. **56**, 854 (1960).

9) *Srivastava, S. N., P. Bahadur*, Colloid & Polymer Sci. **253**, 315 (1975).

10) *Srivastava, S. N.*, J. Indian Chem., Soc. **41**, 203 (1964).

Authors' addresses:

P. M. Gupta
Chemistry Department,
K. R. College, Mathura (India)

P. Bahadur and *S. N. Srivastava*
Chemistry Department,
Agra College,
Agra (India)

Progr. Colloid & Polymer Sci. **63**, 33–40 (1978)

ISSN 0340-255 X

UNESCO Science Centre, Alexandria (Egypt)

The role of non-ionic surfactants in emulsion stability

A. Kamel, V. Sabet, H. Sadek, and *S. N. Srivastava*

With 8 figures and 3 tables

(Received November 29, 1976)

Introduction

A detailed study dealing with non-ionic surfactants, has been given by *Schwartz* and *Perry* (1). The influence of non-ionic surfactants on the quantity and type of low emulsions has been studied by *Matijok* (2). *Elworthy* and *Florence* (3) studied in detail the emulsion stability, using a series of non-ionic surfactants as emulsifiers for emulsions, and have explained their results on the basis of electrical and entropic contributions and the chain lengths of the surfactants used. The effect of some sorbitan esters on the interfacial activity on sodium lauryl sulphate has been observed by *Wan* and *Poon* (4).

The above survey reveals that the data available on non-ionic surfactants are mostly qualitative in nature. Therefore, it was proposed to study quantitatively the stability of emulsions promoted by some non-ionic surfactants in the light of the DLVO theory.

Materials and preparation

The emulsifying agent *W* which is an aryl polyglycol ester, was supplied by Bayer (Germany). The emulsifying agent G 100 P is a mixture consisting mainly of an alkyl phenyl polyglycol ether and a small fraction of an alkylol amide. They were used without further treatment.

The dispense phase of the emulsion was petroleum ether from the Alexandria Petroleum Company, it was distilled, and the 100–120 °C fraction was collected and passed through an alumina column to remove surface active impurities. Pure water was doubly distilled from an all glass apparatus (Barnstead). The inorganic electrolytes and all the other chemicals were obtained from Fischer and used as such.

Dilute o/w G 100P and *W* emulsions were always freshly prepared by suspending 0.2% by volume

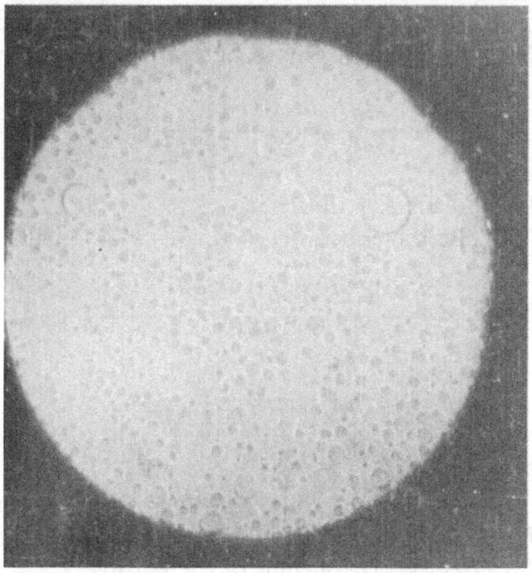

Fig. 1a. Microphotograph of the *W* emulsion

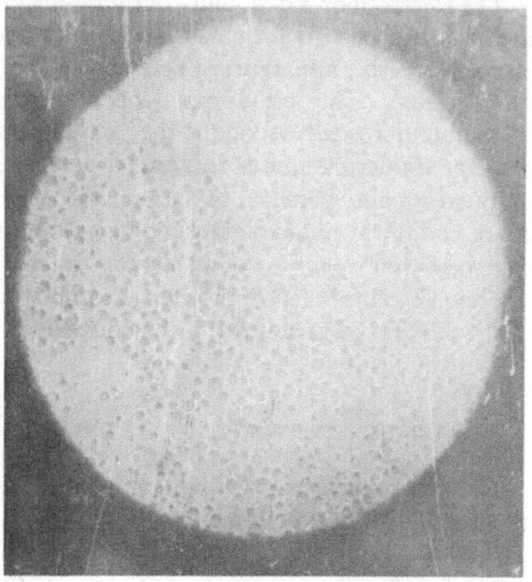

Fig. 1b. Microphotograph of the G 100P emulsion

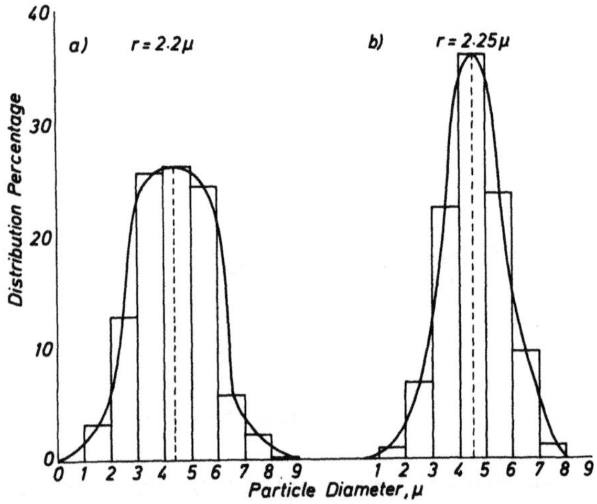

Fig. 2. a) Size-distribution curve for the *W* emulsion, b) Size-distribution curve for the G 100P emulsion

petroleum ether in aqueous solution of 0.0002% of emulsifying agent and vigorously stirred using a high speed homogenizer (Virtis Co., Gardiner, NY, USA).

Experimental and theoretical

The particle size of the emulsions was determined from microphotographs, figures 1a and 1b, where the overall magnifications amounted to 935*x*. For each emulsion the diameter of 400 droplets was measured, thus allowing a size frequency curve to be drawn, figures 2a and 2b, from which the average radius of the *W* emulsion droplets was found to be 2.20 μm and that of G100P to be 2.25 μm.

The electrophoretic mobility of the emulsion droplets has been determined using a microelectrophoresis apparatus (Zetameter Inc., NY, USA). The emulsion droplets were tracked under a microscope as they migrated a distance *d* microns under an applied potential gradient *v* volts, between two electrodes 1 cm apart, and the average migration time *t* seconds was measured by means of an electric timer.

The absolute electrophoretic mobility *u* could thus be calculated from the equation:

$$u = (dt)/(vl) \text{ - - - - - (micron/sec. per volt/cm)}. \quad [1]$$

The zeta potential in millivolts was obtained by substitution in the equation:

$$ZP = 113000 \frac{\eta}{\varepsilon} u \text{ - - - - - (m.v.)} \quad [2]$$

where η and ε are the viscosity in poises and the dielectric constant of the suspended liquid respectively. The values of the conversion factor were obtained by interpolation from a large scale plot as a function of temperature.

The interactions between the colloidal particles of lyophobic systems have been quantitatively treated by *Derjaguin* and *Landau* (5), *Verwey* and *Overbeek* (6) (DLVO theory), on the basis of the free energy of the double layer and the London – Van der Waals' dispersion forces. According to their theory the total interaction energy on mutual approach of two particles is given by:

$$V_{\text{total}} = V_R + V_A \quad [3]$$

where V_R is the repulsive energy and V_A is the attractive energy.

The repulsive energy calculations, between two spherical particles, were carried out by an approximate expression by *Derjaguin* and *Kussakov* (7).

$$V_R = \frac{1}{2}\, \varepsilon a \psi_0^2 \ln (1 + e^{-\varkappa H}) \quad [4]$$

where ψ_0 is the surface potential (assumed to be equal to the zeta potential), ε is the dielectric constant in the double layer, *a* the particle radius, *H* the interparticle distance and $\varkappa$ is the Debye-Hückel parameter.

The attraction energies were calculated, keeping in view the retardation correction, by using the approximate equation due to *Schenkel* and *Kitchener* (8). Two expressions are required for two different ranges of *H*. The equation:

$$VA = -\frac{Aa}{12\,H}\left(\frac{\lambda}{\lambda + 3.54\,H\pi}\right) \quad [5a]$$

is valid for $H < 150$ Å.

While the equation:

$$VA = -\frac{Aa}{\pi}\left(\frac{2.45\,\lambda}{120\,H^2} - \frac{\lambda^2}{1045\,H^3} + \frac{\lambda^3}{5.62 \times 10^4\,H^4}\right) \quad [5b]$$

is valid for $H > 150$ Å.

Where λ is the wave length of the London frequency, taken as 1000 Å and *A* is the Van der Waals constant of the particle.

Thus the interaction energy *V* could be

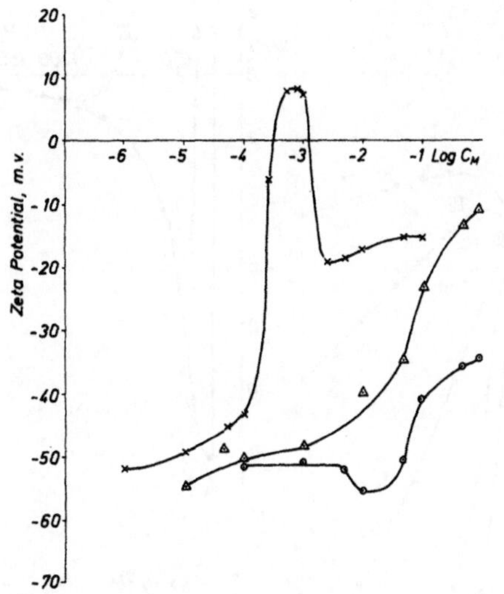

Fig. 3. Curves showing the variation of zeta potential for the *W* emulsion with concentration of electrolytes; ⊙ KNO_3, △ $Sr(NO_3)_2$, × $Al_2(SO_4)_3(NH_4)_2SO_4$

evaluated and an energy profile could be drawn.

It has been pointed out (6), that, knowing *V* – the depth of the potential energy secondary minimum – the number of droplets in a shell *ds* around a central droplet i.e. the degree of aggregation (*D*), can be calculated (9) theoretically from the following equation:

$$D = 4\pi a^3 n_0 \int S^2 e^{-V/KT} ds \qquad [6]$$

where n_0 is the number of monomers / unit volume of the emulsion at equilibrium, and $S = 2 + H/a$.

The integral of the above expression is evaluated graphically, the question of the integration limits has been discussed elsewhere (10).

The rate of flocculation of the emulsions has been studied hemocytometrically, by counting the average number of monomers and dimers per unit volume with time, at different concentrations of electrolytes. Use was made of an optical microscope (American Optical Co., Buffalo, NY, USA), and for each determination about 400 droplets were counted.

In case of reversible or secondary minimum flocculation, a monomer – dimer equilibrium temporarily exists, and the extent of flocculation can be calculated from the equation:

$$D = \frac{\text{No. of dimers}}{\text{No. of monomers} + \text{No. of dimers}} \qquad [7]$$

where *D* is the observed degree of aggregation.

The pH measurements, at different concentrations of electrolytes, were carried out using a Radiometer pH meter, type pH M 28. Some precautions were taken during the measurements as a result of the continuous change of surface properties with time.

Table 1. Influence of electrolyte on zeta potential and energy profiles

Electrolyte conc., *M*	ζ Potential *mV*	Height of max. *kT*	Depth of sec. min., *kT*	Potential *mV*	Height of max. *kT*	Depth of sec min., *kT*
KNO_3	G 100P emulsion			*W* emulsion		
1.0×10^{-5}	50	3972	–	–	–	–
1.0×10^{-4}	51	3751	–	52	4033	–
1.0×10^{-3}	53	3503	0.8	52	3654	0.2
5.0×10^{-3}	58	3360	5.8	52	2953	1.3
1.0×10^{-2}	59	2909	12.1	55	2861	2.9
$Sr(NO_3)_2$						
1.0×10^{-5}	55	4570	–	54	4610	–
5.0×10^{-5}	56	4438	–	49	3536	–
1.0×10^{-4}	56	4294	0.3	50	3688	0.1
1.0×10^{-3}	57	3552	3.4	48	2781	0.7
1.0×10^{-2}	–	–	–	40	870	11.8
$AlNH_4(SO_4)_2$						
1.0×10^{-6}	–	–	–	52	4158	–
1.0×10^{-5}	54	4093	–	49	3596	–
5.0×10^{-5}	–	–	–	45	2701	0.2
1.0×10^{-4}	54	3452	1.7	43	2338	0.5
2.5×10^{-4}	36	990	6.9	–	–	–

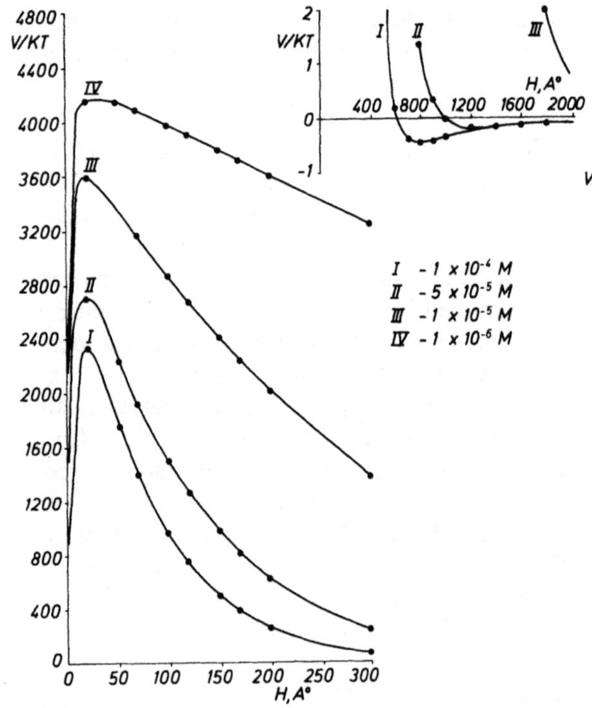

Fig. 4. Potential energy profiles and elaborated secondary minima for the *W* emulsion flocculated by $AlNH_4(SO_4)_2$

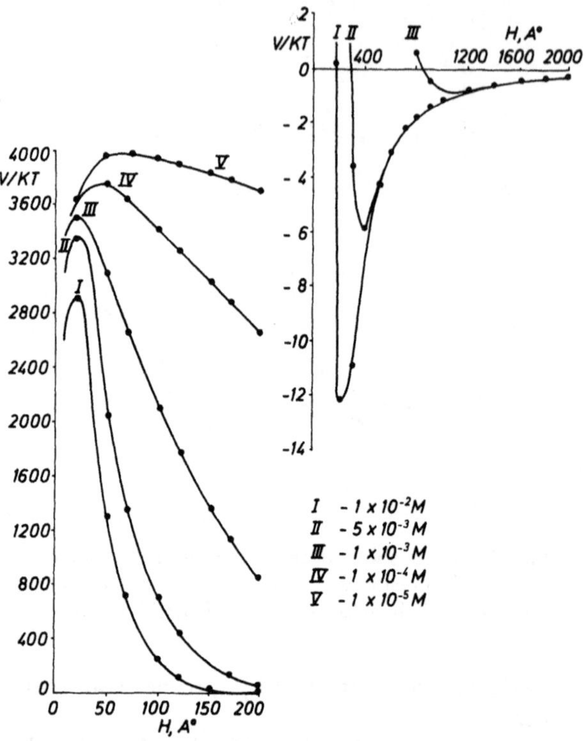

Fig. 5. Potential energy profiles and elaborated secondary minima for the G 100P emulsion flocculated by KNO_3

Results and discussion

The influence of electrolytes on the zeta potential of the emulsion droplets

The observed values of zeta potentials of the initial emulsions were -56 and -55 m. v. for the *W* and G 100P emulsions respectively, at room temperature (23 $\pm$2 °C). The flocculation was, therefore, brought about by different valent cations. The general nature of the effect of electrolytes, viz. KNO_3, $Sr(NO_3)_2$ and $AlNH_4(SO_4)_2$ on the zeta potentials of the emulsions has been shown through z.p. – log *C* plots as represented in figure 3 for the *W* emulsion. The plot for the G 100P emulsion was similar in nature.

In general, the zeta potential decreases as the concentration of electrolytes increases, this effect has usually been explained in terms of the specific adsorption of counter ions into the Stern layer.

KNO_3, on the contrary, first showed an increase in zeta potential for both emulsions, this was partly attributed to a decrease in dielectric constant (11, 12) and partly to a specific stabilizing action of adsorbed indifferent nitrate ions (13, 14).

From figure 3, it is obvious that the Al^{+3} ion reverses the sign of the charge, this is usually observed with highly charged or strongly adsorbable counter ions. This behaviour was explained by assuming that Al^{+3} ion is very easily hydrolized and the products of hydrolysis, which presumably are positively charged oxides or hydroxides, are able to reverse the sign of the charge (15).

Interaction energy profiles and stability of the emulsions

For the assessment of the stability of both emulsions the interaction energy has been calculated from eq. [3–5]. The interaction energies between the surfaces of two droplets have been plotted as a function of distance in the presence of different electrolytes, as represented in figures 4, 5. The nature of the curves is the same for other electrolytes also. The relevant data are recorded in table 1.

The interaction energies obtained for both emulsions, as represented in the profiles, are of very high energy barriers, 25 kT, and thus the systems are highly stable, and the possibility of occurrence of flocculation in the primary minima is completely excluded. The height of the energy barriers varies with ζ and $\varkappa$, which depend on the electrolyte content and valency. The presence of electrolytes causes a compression of the diffuse double layer around the particles and may, in addition, exert a specific effect through ion adsorption into the Stern layer leading to a decrease in ζ, this behaviour increases with the increase in valency.

It is obvious from table 1 that both emulsions are of comparatively equal stability, the G 100P emulsion may show greater stability in the higher range electrolyte concentrations, while the *W* emulsion shows this stability in the lower range.

From the plots of the elaborated secondary minima, shown in figures 4, 5, it is clear that the minima range from 1–12 kT which is reasonable for the occurrence of flocculation. As the attraction term is proportional to the droplet radius, eq. [5], it is obvious that the secondary minima in both systems should become significant, as the radii of both emulsions are considerably large. Evidence already exists that this type of flocculation occurs in the suspensions of relatively large particles (16, 17). It was concluded that the high energy values and the occurrence of secondary minima in the *W* and G 100P emulsions are generally due to their large particle size.

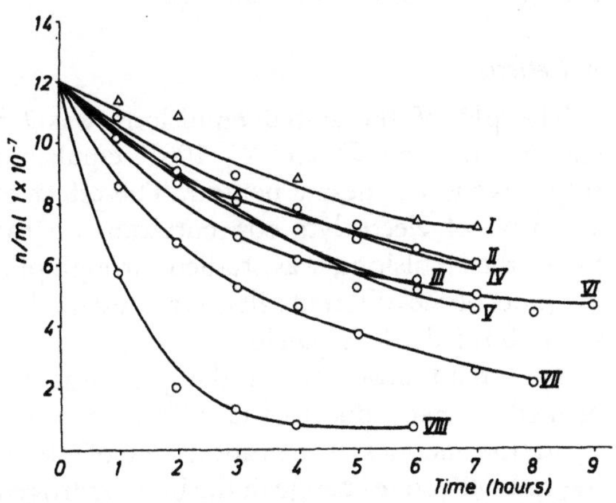

Fig. 6. Curves showing the change in number of monomers (n) with time for the G 100P emulsion. I - zero conc., II - 1×10^{-3} M KNO_3, III - 1×10^{-2} M KNO_3, IV - 1×10^{-3} M $Sr(NO_3)_2$, V - 5×10^{-2} M $Sr(NO_3)_2$, VI - 1×10^{-4} M $AlNH_4(SO_4)_2$, VII - 7.5×10^{-4} M $AlNH_4(SO_4)_2$, VIII - 5×10^{-3} M $AlNH_4(SO_4)_2$

Degree of aggregation and Van der Waals constant of the emulsion droplets

The observed degrees of aggregation (D) were obtained, by counting the numbers of individual oil droplets (monomers) against time, these monomers were then plotted against time, as represented by figure 6 for the G 100P emulsion. The curves are observed to lower down by the increase of the flocculating electrolyte concentrations, showing the coagulating effect of electrolytes on the emulsion. The curves also seem to level off after some time, indicating the reversible and equilibrium nature of the flocculation process. At equilibrium, D's were calculated from eq. [7], and the relevant data for the G 100P emulsion are recorded in table 2.

The interaction energy profiles of both emulsions were calculated using various values of Van der Waals constant (A), for each ζ. For each curve, with a given value A of the theoretical degree of aggregation was obtained by graphical integration of eq. [6]. These degrees of aggregation were then plotted against Van der Waals constant A, as shown in figure 7 for the G 100P emulsion and tabulated in table 2. The value of A corresponding to the experimentally determined D was then read off for each system, and the mean value determined.

The mean effective values of A for petroleum ether coated particles were found to be equal to 1×10^{-13} ergs for the *W* emulsion and 3×10^{-13} ergs for the G 100P emulsion. These values are somewhat higher than the calculated value of 7×10^{-14} ergs obtained from physico-optical data (18). This was assumed to be due to the adsorbed film of stabilizing macromolecules (19). The greater value of A for the G 100P emulsion over that of the *W* emulsion, is presumably due to the thicker sheath of the G 100P emulsifier formed of an alkyl phenyl polyglycol ether with an alkylol amide, while the *W* emulsifier consisting of an aryl polyglycol ester is expected to have a thinner adsorbed interfacial layer, and consequently a lower A.

The values of D obtained were very low, as

Table 2. A's as a function of degree of aggregation

Electrolyte conc., M	Van der Waals constant (A) $\times 10^{13}$	Theoretical D%	Experimental D%	Corresponding A $\times 10^{13}$
KNO_3				
1×10^{-3}	1.0	0.04		
	5.0	0.12	0.01	4.0
	10.0	0.47		
1×10^{-2}	0.5	0.07		
	1.0	0.13	1.55	2.0
	3.0	8.70		
$Sr(NO_3)_2$				
1×10^{-3}	1.0	0.07		
	4.0	0.64	0.05	3.0
	5.0	1.91		
$AlNH_4(SO_4)_2$				
1×10^{-4}	1.0	0.05		
	5.0	0.23	0.02	4.0
	10.0	9.61		

Average value of $A = 3.30 \pm 0.62 \times 10^{-13}$

seen from table 2, this may be attributed to the fact that as soon as flocculation occurs, coalescence immediately proceeds, this is probably due to the weakness of the interfacial film, having a low concentration of 2×10^{-6} V/V ratio.

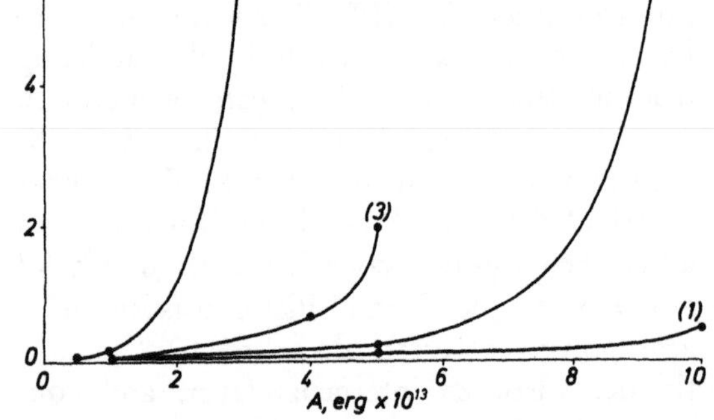

Fig. 7. Degrees of flocculation as a function of Van der Waals constant for the G 100P emulsion

Stabilization by steric hindrance

Stabilization due to steric hindrance is also likely to occur in both systems, as a result of the bulkiness of the non-ionic surfactants used. Polyoxyethylene glycols were reported to be efficient steric stabilizers, at the same time alkalonamides have proved to be excellent detergent adjuncts to non-ionic surfactants (20).

The steric stabilization induced by these bulky surfactants has been explained by the theory of entropy stabilization (21).

pH effects

The pH of the initial emulsions was 6.5 and 6.6 for the W and G 100P emulsions respectively, i.e. nearly neutral. The effect of variation of electrolyte concentration on the pH of the emulsions, was studied through pH-log plots of the different salts, as represented by figure 8 for the W emulsion.

The trend observed in the plot was explained as being due to the difference in the hydration energies of the cations, table 3, leading to an effective change in the bulk hydrogen ion concentration.

Table 3, variation of the hydration energy with the cations of the electrolytes used.

The striking similarity of the pH-log C plots of both systems indicates the independence of pH on the nature of the non-ionic

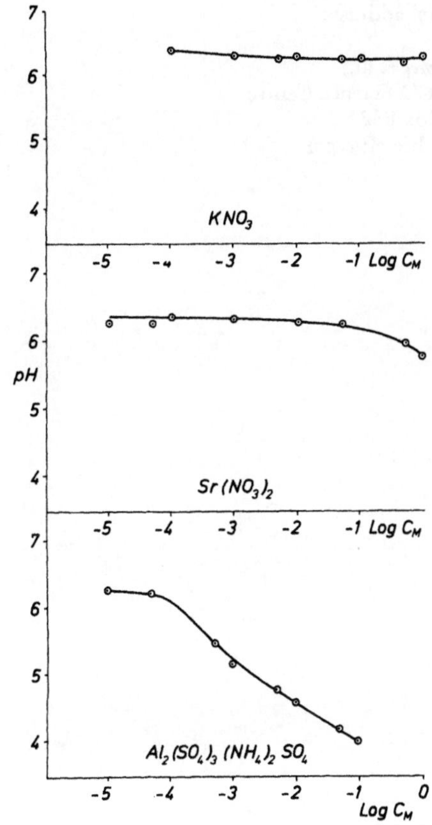

Fig. 8. Curves showing the change in pH of the *W* emulsion with log molar concentration of electrolyte

Table 3. See text

Cation	K^+	Sr^{+2}	Al^{+3}
H (k. cal/mole)	77	355	very high

surfactant used, this strengthens the conclusion that non-ionic stabilized emulsions are not affected by pH variations (20).

Summary

The role of *W*, aryl polyglycol ester, and G 100P, an alkyl phenyl polyglycol ether, on the stability of dilute petroleum ether in water emulsion has been studied in the light of the DLVO theory. The zeta potential of the emulsions flocculated by varying concentrations of different electrolytes, viz., potassium nitrate, strontium nitrate and aluminium ammonium sulphate, has been calculated from the mobility data obtained by a microelectrophoretic technique. With the help of these data, the interaction energy profiles were calculated by summing up the attraction and repulsion energies. The high energy barriers so obtained manifested the stability of both systems. The theoretical degrees of aggregation were calculated for varying Van der Waals constant.

The electrolyte flocculation was studied haemocytometrically by counting the number of associated and unassociated droplets, and the degrees of aggregation were calculated. By interpolating the observed degree of aggregation into the theoretical curves the operating Van der Waals constant of both systems was obtained, and found to be 1×10^{-13} erg and 3×10^{-13} erg for the *W* and G 100P systems respectively. Stabilization due to steric hindrance was also suggested to occur in these systems and was explained by the theory of entropy stabilization. The pH measurements were carried out, and it was found that emulsions stabilized by non-ionic surfactants are not pH sensitive.

Zusammenfassung

Es wurde die Wirkung von Arylpolyglycolester (*W* der Bayer AG) und Alkylphenylpolyglycoläther (G 100P) auf die Stabilität verdünnter Petroläther-in-Wasser-Emulsionen im Licht der DLVO-Theorie untersucht. Die Zetapotentiale wurden aus mikroelektrophoretischen Daten ermittelt. Außer der Flockung durch Elektrolyte wurden Hinweise auf eine Stabilisierung durch sterische Hinderung gefunden, die im Sinne einer Entropiestabilisierung gedeutet wurden.

References

1) *Schwartz, A. M., J. W. Perry*, Surface Active Agents, p.p. 15, 17 (New York 1949).

2) *Matijok*, Ann. Farmacia (Bucharest) **18**, 107 (1970).

3) *Elworthy, P. H., A. T. Florence*, J. Pharm. Pharmacol. **21**, 705 (1969).

4) *Wan, Poon*, Can. J. Pharm. Sci. **6**, 59 (1971).

5) *Derjaguin, B., Landau*, Acta Phys. Chim. **14**, 633 (1941).

6) *Verwey, E. J. W., J. T. G. Overbeek*, Theory of the Stability of Lyophobic Colloids (Amsterdam 1948).

7) *Derjaguin, B., M. Kussakov*, Acta Phys. Chim. **10**, 25, 153 (1939).

8) *Schenkel, J. H., J. A. Kitchener*, Trans. Faraday Soc. **56**, 161 (1960).

9) *Prakash, C., S. N. Srivastava*, Bull. Chem. Soc. Japan **40**, 1765 (1967).

10) *Srivastava, S. N., D. A. Haydon*, Trans. Faraday Soc. **60**, 971 (1964).

11) Colloid Symposium Monograph **4**, 304 (1926).

12) *Kruyt, H. R., P. C. van der Willigen*, Z. Phys. Chem. **130**, 170 (1927).

13) *Mukerjee, J. N.*, J. Indian Chem. Soc. **5**, 735 (1928); **7**, 465 (1930); **10**, 26 (1933); Nature **122**, 960 (1928); Science, Culture **I, III** (1935).

14) *Ghosh, B. N.*, J. Chem. Soc., 2693 (1929).

15) *Trolestra, S. A.*, Thesis, p.p. 125, 132 (Utrecht 1941); *H. R. Kruyt, S. A. Trolestra*, Kolloid Beihefte **54**, 277, 284.

16) *Schenkel, J. H., J. A. Kitchener*, Trans. Faraday Soc. **56**, 161 (1960).

17) *Srivastava, S. N., D. A. Haydon*, Trans. Faraday Soc. **60**, 971 (1964).

18) *Prakash, C., S. N. Srivastava*, Z. Phys. Chem. **244**, 200 (1970).
19) *Vold, M. J.*, J. Coll. Sci. **16**, 1 (1961).
20) *Schick, M. J.*, Non Ionic Surfactants (New York 1967).
21) *Mackor, E. L.*, J. Coll. Sci. **6**, 492 (1951).

Authors' address:

A. Kamel et al.,
UNESCO Science Centre
P.O. Box 832
Alexandria (Egypt)

Progr. Colloid & Polymer Sci. **63**, 41–49 (1978)

ISSN 0340-255 X

Agra College, Agra (India)

Role of finely dispersed solids (hydroxides of polyvalent metals) as emulsifiers

S. N. Srivastava

With 8 figures and 7 tables

(Received December 6, 1976)

Introduction

The first attempt to explain the stability of solid stabilized emulsions in a quantitative manner was made by *Van der Minne* (1) who showed that under thermodynamic equilibrium the adsorption of solid particles at the liquid – liquid interface is governed by its contact angle against the solid, an obtuse contact angle against the solid phase facilitating the stabilization. All subsequent theories on solid stabilised emulsions attributed their stability to the effect of the contact angle. In this connection the work of *Schulman* and *Leja* (2) is noteworthy. *Albers* (3) quantitative formulations of the stability of emulsions of water-in-benzene with basic carbonates and hydroxides of polyvalent metals formed by the hydrolysis of corresponding oleates have also received attention. Other pertinent references are those of the work of *King* and *Mukerjee* (4), *Lucassen-Reynders* (5) and *Vold* (6). The latest elaborate work on the subject is that of *Mukerjee* and *Srivastava* (7–12) and *Jain* and *Srivastava* (13).

From the survey of the solids used as emulsifying agent it will be seen that the property of emulsification is possessed by a large number of them but owing to coarseness and instability their use is rather limited. However, they find extensive applications as paints, varnishes and cosmetics. Nevertheless, quantitative investigations on solid emulgents are rather sparse. The present study is an attempt in this direction and has been conducted on oil-in-water emulsions stabilised by a variety of hydroxides of polyvalent metals. Emulsion droplets coated with adsorbed metal hydroxide films exhibit appreciable stability both to flocculation and coalescence under normal conditions. They interact with electrolytes, complex ions and surfactants as if they were hydrophobic colloids. Investigations on these aspects of the emulsion behaviour are also described.

Theoretical

A. Rate of flocculation

From Smoluchowski's (14) theory we have:

$$\frac{1}{n} - \frac{1}{n_0} = 4\pi\mathcal{D}Rt\,. \qquad [1]$$

Where n_0 and n are the number of singlets present initially and at time t. $\mathcal{D}$ is the diffusion coefficient and $R = 2a$, a being particle radius. If φ is the phase volume (if the oil used is 5% $\varphi = \frac{5}{100} = 0.05$) of the O/W emulsion and n the No. of drops of mean volume Vm, after time t, we have by definition:

$$Vm = \varphi/n. \qquad [2]$$

Combining [1] and [2] we have:

$$Vm = \varphi/n_0 + 4\pi\mathcal{D}R\varphi t. \qquad [3]$$

Because the phase volume remains unchanged during coalescence we have also:

$$\varphi = nV_m = n_0 Vm_0$$

where Vm_0 is the drop volume at $t = 0$. Hence

$$Vm = Vm_0 + 4\pi\mathcal{D}R\varphi t = K_1 t \qquad [4]$$

where K_1 is the Smoluchowski's rate constant for flocculation.

Again it will be seen that these emulsions stabilised by the metal hydroxides have a much higher $t_{1/2}$ than predicted by Smoluchowski's theory ($t_{1/2} = 2 \times 10^{11}/n_0$). This implies that they have an energy barrier represented by a stability factor W_1. So [4] may be modified as:

$$Vm = Vm_0 + 4\pi\mathcal{D}R\varphi t\, e^{-W1/kT}\,. \quad [5]$$

Plot of Vm against t will be a straight line. The slope of this curve $= 4\pi\mathcal{D}R\varphi e^{-W1/kT}$.
So making use of the Einstein equation $\mathcal{D} = kT/6\pi\eta\, a$ one can calculate W_1 which should be of several kT for stable emulsions.

B. Rate of coalescence

The rate of change of specific interface (interfacial area per gm of the dispersed liquid) is proportional to the interfacial area and is given by the equation:

$$-\frac{ds}{dt} = \frac{S}{K} = K_2S\,. \quad [6]$$

Where $K_2 = \frac{1}{K}$, being the stability factor and is defined as the reciprocal of the rate of decrease of each *sq*, cm. of fresh emulsion per unit time.

$$\text{or}\ \ln S - \ln S_0 = -K_2t \quad [7]$$

$$\text{or}\ S = S_0e^{-K_2t} \quad [8]$$

$$\text{or}\ \log S_0 - \log S = K_2\ 1/2.303\,. \quad [9]$$

This equation conforms to the first order kinetics and so coalescence is kinetically of first order. Plot of ln S against t should give a St line whose slope will give K_2, the rate of coalescence of the investigated systems.

To calculate the stability factor the slope of each curve has been measured and divided by the initial specific area of interface to give the rate of decrease per unit aıea of interface; this is termed the instability factor $\left(\frac{ds}{dt}\right)$ and is defined as decrease of each Sq. cm. of fresh emulsion per day. The reciprocal of this is a direct measure of emulsion stability.

C. Application of the DLVO theory

In the light of the DLVO theory (15) the interaction energy of the hydroxide coated emulsion droplets is given by

$$V = V_R + V_A \quad [10]$$

where V_R and V_A are respectively the repulsion and the attraction energies and are respectively given by equations [11] and [12] as given below

$$V_R = \frac{\varepsilon a \psi_0^2}{2} \ln\left(1 + e^{-KH}\right). \quad [11]$$

Where V_R is the energy of repulsion, ε, dielectric constant, a, the particle radius, K the Debye Hückel parameter and H the interparticle distance. ψ_0 is the surface potential assumed to be the zeta potential for the present systems having $Ka \gg 1$.

$$V_A = -\frac{Aa}{\pi}\left(\frac{2.45\,\lambda}{120\,H^2} - \frac{\lambda^2}{1045\,H^3} + \frac{\lambda^3}{5.62 \times 10^4\,H^4}\right). \quad [12a]$$

Valid for $H > 150$ Å,

$$\text{and}\ V_A = -\frac{Aa}{12\,\pi}\left(\frac{\lambda}{\lambda + 3.54\,\pi\,H}\right) \quad [12b]$$

Valid for $H < 150$ Å.
λ is the London's wave length. The value of A used here is 2×10^{-12} ergs (15). The following approximate formula due to *Derjaguin* (16) was also used

$$P = \frac{\varepsilon a \psi^2}{2\,kT} \quad [13]$$

where P should be greater than 10 for stable systems. Such a formulation will be only to the first approximation as Van der Waals interactions have not been taken into account.

Eilers empirical formulations

Eilers and *Korff* (17) demonstrated that the factor which predominates the stability of a colloidal system is the energy required to bring together two particles of the dispersion against the action of their electric field. The following formula may be derived for this energy E:

$$E = \xi^2/\varkappa \quad [14]$$

where ξ is in mV and the Debye-Hückel reciprocal distance, K in cm^{-1}. More generally,

Table 1. Initial characteristics of the hydroxides stabilised emulsions

Emulsion No.	Name of emulgent	Average particle size of the disperse hydroxide (μm)	pH of the emulsion	ζ (mv)	a (μm) for emulsion droplets	$\varkappa \times 10^{-7}$	Height of the energy barrier in kT ($\times 10^{-3}$)	$\frac{\varepsilon a \psi^2}{2\,kT}$ ($\times 10^{-3}$) in kT
1.	$Al(OH)_3$	0.21	4.1	32	2.35	0.04	1.7	2.7
2.	$Cd(OH)_2$	0.36	7.6	20	2.75	2.13	0.3	1.1
3.	$Cr(OH)_3$	0.23	5.3	30	2.38	0.28	1.3	2.4
4.	$Fe(OH)_3$	0.26	5.5	40	2.58	0.01	2.9	4.6
5.	$Mg(OH)_2$	0.37	10.5	17	2.86	0.59	0.4	9.2
6.	$Co(OH)_2$	0.27	6.8	23	2.72	1.09	0.6	1.6
7.	$Zn(OH)_2$	0.24	7.1	28	2.39	0.65	1.0	2.1
8.	$Ce(OH)_3$	0.31	7.4	18	2.68	2.69	0.2	9.7
9.	$Th(OH)_4$	0.25	3.5	36	2.62	0.04	2.4	3.8
10.	$Ti(OH)_3$	0.33	3.2	45	2.76	0.01	3.9	6.2
11.	$La(OH)_3$	0.30	8.4	18	2.82	3.69	0.1	1.0

$$C\xi\left(\frac{\varepsilon}{\varkappa}\right)^{1/2} = \text{constant} \qquad [15]$$

or in rigorous theoretical form (18) taking into account the Van der Waals constant, A and replacing ξ by ψ_0 one has for stable systems

$$\frac{2\varepsilon\psi^2}{\varkappa A} > 1\,. \qquad [16]$$

To calculate zeta potential, ζ, from electrophoretic mobilities (4) the following equation due to Helmholtz-Smoluchowski was used.

$$= \frac{4\pi\eta}{\varepsilon X}\cdot u$$

where η is the viscosity of the medium, X is the field strength. The equation is applicable for $\varkappa a \gg 1$ and charge densities $< 1.4 \times 10^4$ esu per cm^2 which is actually the case with the present.

Materials and methods

The hydroxides/hydrous oxides of the polyvalent metals used as emulgents were freshly prepared by the method reported earlier (7). Analar petroleum ether (B.D.H.) having a boiling range of 120–160 °C and a density of 0.756 g cm^{-3} was used after purification through alumina column. The Triton X non-ionic surfactants were from Rohm and Haas U.S.A. whereas the cationic surfactants were from B.D.H. Other materials were of A.R. grade and were used as such.

The emulsions were prepared by dispersing 1.0% by volume of petroleum ether in 1% aqueous suspension of the corresponding hydroxide. The mixture was then homogenized in a Fischer homogenizer with Stainless Steel stirrer and Pyrex glass container. The emulsions were then sampled under identical conditions.

The flocculation of the emulsion was followed haemocytometrically using an improved Neubauer model of hemocytometer and Fischer Tally counter under Leitz microscope with 15 × 95 magnification. The pH adjustments were done by a Cambridge pH meter. The zeta potentials were calculated from electrophoretic mobilities which were determined in a rectangular Northrop and Kunitz flat type micro-electrophoretic cell attached to a Carl Zeiss Jena microscope, the details being described elsewhere (13).

Results and discussion

Hydroxides/hydrous oxides of some 11 metals were used to stabilize petroleum ether in water emulsions whose initial characteristics are listed in table 1. These emulsions although rather coarse, the droplet size being greater than 1 μm are fairly stable. At first there was slightly rapid deterioration but soon the emulsions stabilized themselves without oil separation for one month. Their high stability supports the view that the emulsion stability is unrelated to the degree of dispersion. Like protein (19) (bovine serum albumin) stabilized emulsions their zeta potentials are rather low and range from 18 to 45 mV. However, they are not appreciably pH dependent.

Flocculation was examined by analysing the droplet concentration data in the framework of the Smoluchowski's theory and its modified form as presented in the theoretical portion. Firstly the droplet concentration n was plotted against time as denoted in figure 1 wherein the

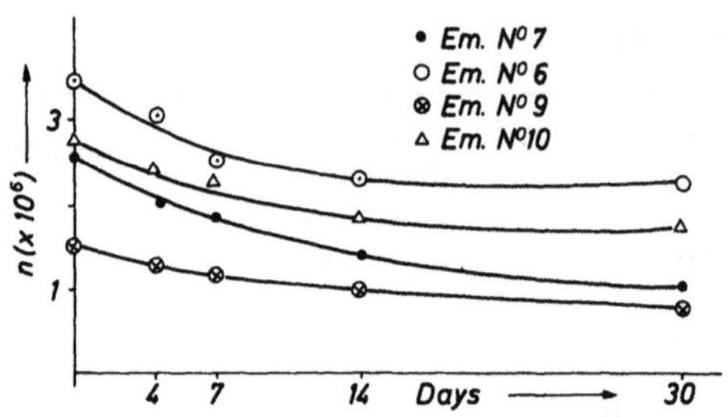

Fig. 1. Plot of n vs. t

curves show an eventual levelling off owing to the existence of a less degenerate quasi equilibrium state assumed to arise from the coexistence of monomers and dimers. This was due, in all probability, to reversible flocculation occurring in the secondary minima of energy profiles as discussed below. One could further check this by gently shaking the emulsion after a reasonable time and noticing that the droplets were deflocculated again and the equilibrium was once more disturbed.

Thus when $\frac{1}{n}$ (Cf. equation 1) was plotted against t in figure 2, it was found that the value of the flocculation rate constant k is extremely low of the order of 10^{-15} cm^3 sec^{-1}. This may be assigned to the energy barriers which inhibit the flocculation and change even its nature. The energy barriers, W_1 can be assessed from the plots of Vm against t represented in figure 3. It was found that the energy barriers to flocculation are of the order of 15 kT or more and so the flocculation is likely to be extremely slow and not of irreversible nature.

Using the equations [11] and [12a] of the

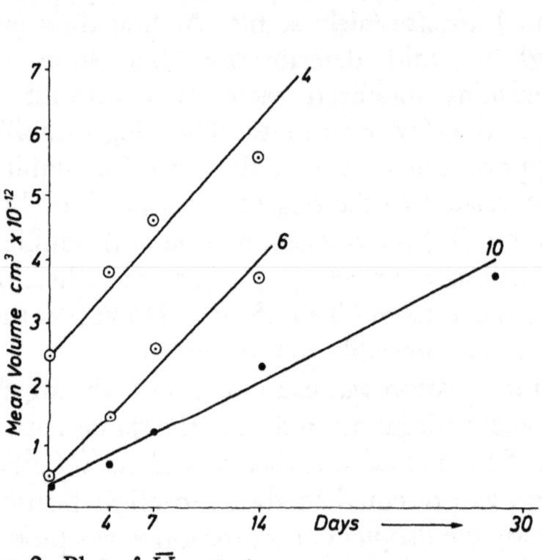

Fig. 2. Plot of $\bar{V}m$ vs. t

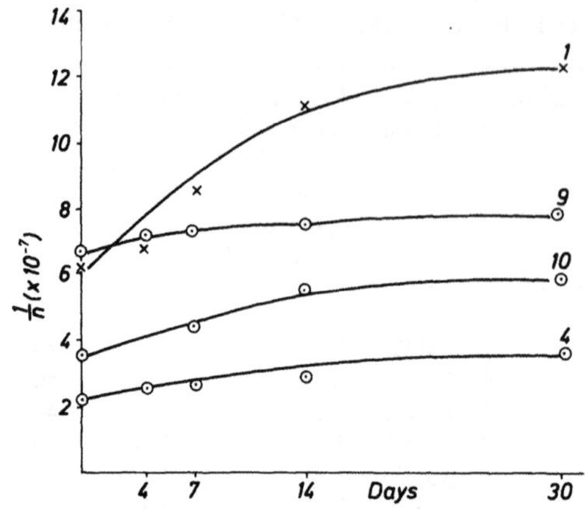

Fig. 3. Plot of $\frac{n}{l}$ vs. t

DLVO theory the interaction energy profiles for the original emulsions were constructed. They are all very similar having a high potential energy barrier (value of the Van der Waals constant being 12×10^{-12} erg) as listed in the penultimate column of table 1. This reflects that the flocculation cannot occur except in the secondary minimum of the energy profiles as denoted in a typical curve of figure 4 wherein, the energy minimum is deep enough

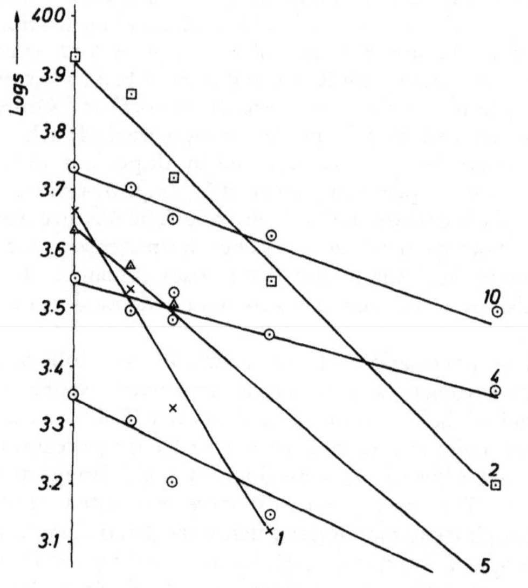

Fig. 4. Plot of log s against t

($\approx 6\,kT$) to trap the particles and thereby lead to flocculation. So the experimental flocculation seems to be consistent with the theoretically predicted flocculation based upon the zeta potential measurements.

For the calculation of stability an approximate formula (eqn. [13]) was used and the results are given in the last column of table 1. All values are of the order of $10^3\,kT$ which is much greater than $10\,kT$, the condition derived by *Derjaguin* for stable dispersions. In general, these values are higher than the corresponding values of the penultimate column because in the approximate formula, the effect of the Van der Waals interaction is not considered.

However, when the electrolytes $Pb(NO_3)_2$ and $UO_2(NO_3)_2$ are added the zeta potential decreases with continued addition and flocculation sets in when the added salt compresses the double layer and reduces zeta to a point where the repulsive forces can no longer prevent the surfaces of emulsion droplets coming into intimate contact. This happens when the value of the factor $\zeta^2/\varkappa$ decreases below 10^{-3}, a condition derived by *Eilers* and to the present systems as revealed from tables 2 and 3. Though the function $\zeta^2/\varkappa$ is a measure of emulsion stability quantitative agreement is often unsatisfactory if there is appreciable long range attraction. Under such circumstances the stability criterion for weakly charged particles is given by the expression (16) which is tabulated in the penultimate columns of tables 2 and 3. It will be seen that except for the first values of 1.61&1.75 respectively with $Pb(NO_3)_2$ and $UO_2(NO_3)_2$ the rest are lower than unity and hence are indicative of onset of instability. In the last column of these two tables the modified Eilers' formula is found to fit better and yield more or less constant values.

Thus, as discussed above, owing to the presence of high potential energy barriers, the flocculation rate constant of original emulsions is of the order of 10^{-15} cm^3 see^{-1} and about 10^3 times smaller than the theoretical Smoluchowski's constant (Cf. table 4). However, in relation to coalescence the flocculation is faster and hence as discussed below the former is the rate determining step.

Coalescence

Equations [7 to 9] have been used to calculate the rate of coalescence with the help of the plots of figure 5. The rate is of the order of 10^{-4} see^{-1} which is much slower than that of the macromolecular (20) stabilized emulsion

Table 2. Addition of $Pb(NO_3)_2$ on $Th(OH)_2$ stabilized emulsion (if $\zeta^2/\varkappa$ is less than 10^{-3}, coagulation will set in)

ζ	$C\,(M)$	$\varkappa$	$\zeta^2/\varkappa$ (for ζ in mV)	$2\varepsilon\zeta^2/A\varkappa$	$C\zeta\left(\frac{\varepsilon}{\varkappa}\right)^{\frac{1}{2}} \times 10^5$
32	1×10^{-4}	0.57×10^6	1.8×10^{-3}	1.61	3.7
25	2×10^{-4}	0.80×10^6	7.5×10^{-4}	0.69	5.0
22	3×10^{-4}	0.98×10^6	4.7×10^{-4}	0.43	5.8
17	4×10^{-4}	1.13×10^6	2.6×10^{-4}	0.21	5.7
15	5×10^{-4}	1.21×10^6	1.7×10^{-4}	0.16	5.9
13	6×10^{-4}	1.39×10^6	1.2×10^{-4}	0.11	6.0

Table 3. Addition of $UO_2(NO_3)_2$ on $Th(OH)_2$ stabilized emulsion

ζ	$C\,(M)$	$\varkappa$	$\zeta^2/\varkappa$	$2\varepsilon\zeta^2/A\varkappa$	$\zeta\left(\frac{\varepsilon}{\varkappa}\right)^{\frac{1}{2}} \times 10^5$
22	1×10^{-4}	0.57×10^6	9.120×10^{-4}	1.75	2.7
17	2×10^{-4}	0.80×10^6	3.785×10^{-4}	0.33	3.4
13	3×10^{-4}	0.98×10^6	1.649×10^{-4}	0.14	3.4
11	4×10^{-4}	1.13×10^6	1.071×10^{-4}	0.09	3.7
10	5×10^{-4}	1.27×10^6	0.827×10^{-4}	0.07	3.6
8	6×10^{-4}	1.39×10^6	4.366×10^{-5}	0.03	3.5

Table 4. Kinetic parameters of the metal hydroxides stabilised emulsions

Emulsion No.	Agent	pH	$K_1 \times 10^{15}$ cm^3 sec^{-1}	$\frac{K_0}{K_1} \times 10^{-3}$	$K_2 \times 10^7$ sec^{-1}	Stability factor (10^{-3}) from interfacial area
1.	$Al(OH)_3$	4.1	3.2	3.1	1.2	3.8
2.	$Cd(OH)_2$	7.6	5.1	1.9	4.6	3.2
3.	$Cr(OH)_3$	5.3	3.4	2.9	1.3	5.6
4.	$Fe(OH)_3$	5.5	2.6	3.8	1.4	5.2
5.	$Mg(OH)_2$	10.5	4.3	2.3	2.4	2.4
6.	$Co(OH)_2$	6.8	5.1	1.9	3.1	4.1
7.	$Zn(OH)_2$	7.1	4.8	2.1	4.2	2.6
8.	$Ce(OH)_3$	7.4	2.0	5.0	1.7	6.2
9.	$Th(OH)_4$	3.5	2.1	4.7	1.5	5.7
10.	$Ti(OH)_3$	3.2	2.9	3.4	2.0	4.7
11.	$La(OH)_3$	8.4	2.5	4.0	1.6	2.4

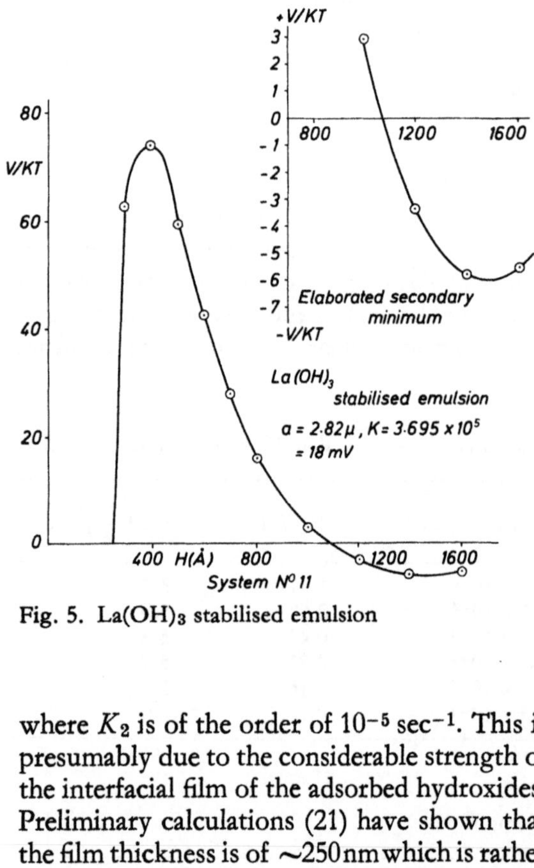

Fig. 5. $La(OH)_3$ stabilised emulsion

where K_2 is of the order of 10^{-5} sec^{-1}. This is presumably due to the considerable strength of the interfacial film of the adsorbed hydroxides. Preliminary calculations (21) have shown that the film thickness is of ~250 nm which is rather of colloidal dimension than molecular one. This implies that clusters of individual hydroxide particles constitute the film. In view of this it is likely that steric hindrance might be preventing coalescence which should perhaps be caused by the removal of solid particles at least from a fraction of two dispersed droplets. The high stability is also demonstrated by the relatively large stability factors (table 4) derived from the rate of decrease of interfacial areas which is rather slow.

To judge whether flocculation or coalescence is the rate determining step ratio of the half lives of these two processes was ascertained. Half life of flocculation as given by equation [1] is $\frac{1}{K_1 n_0}$ whereas that of coalescence is deduced from the equation (Cf. equation [7]) $\ln n_0 - \ln n = k_2 t$ (assuming that it is a first order kinetics is found to be $\frac{1}{1.47 K_2}$). The relative rate is given by the ratio of the two half lives i.e. $\frac{1.47 K_2}{n_0 K_1}$. Flocculation or coalescence is more rapid according as this ratio is greater or less than unity. Since in the present case the ratio is much greater than unity, coalescence is slower and hence determines the rate of coagulation of these emulsions (22).

Effect of complex salts

Recently *Matijevic* and *Kolak* (23) have studied the interaction of AgBr sol with a number of metal (Ni, Co, Cr) chelates which have much lower flocculation value than simple inorganic ions of the same charge. These complex salts have the additional advantage over the more usual Al and Th salts in that they do not form colloidal systems by hydrolytic dissociation. Here the

Table 5. Influence of complex ions (of Co^{III}) on $Th(OH)_3$ stabilised emulsions

Complex ion electrolyte	Zeta in mV					
	Molar concentration					
	5×10^{-6}	1×10^{-5}	2.5×10^{-5}	5×10^{-5}	7.5×10^{-5}	1×10^{-4}
$[Co(NH_3)_4(NO_2)_2]Cl$	35	31	25	18	16	16
$[Co(NH_3)_5Cl]Cl_2$	33	28	24	17	16	15
$[Co(NH_3)_6]Cl_3$	31	27	22	16	14	14
$\left[(NH_3)_4Co\langle{}^{NH_2}_{OH}\rangle Co(NH_3)_4\right]Cl_4 \cdot 4H_2O$	31	26	18	14	13	12
$\left[Co\left\{{}^{OH}_{OH}\rangle Co(NH_3)_4\right\}_3\right]Cl_6$	29	22	17	13	12	12

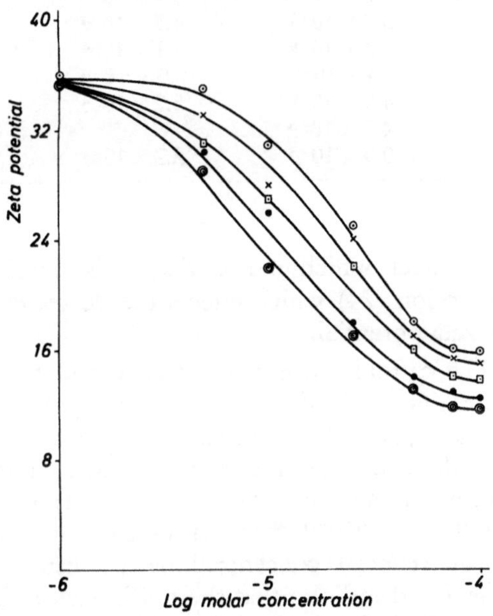

Fig. 6. Influence of complex ions on zeta potential. ⊙ $[Co(NH_3)_4(NO_2)_2]Cl$, × $[Co(NH_3)_5Cl]Cl_2$, □ $[Co(NH_3)_6]Cl_3$, ● $\left[(NH_3)_4Co\langle{}^{NH_2}_{OH}\rangle Co(NH_3)_4\right]Cl_4 \cdot 4H_2O$, ⊙ $\left[Co\left\{{}^{OH}_{OH}\rangle Co(NH_3)_4\right\}_3\right]Cl_6$

complex Co salts with varying complex cations have been used to coagulate the negatively charged emulsions of the present system. It can be seen from table 5 that with complex ions relatively lower concentrations are required to bring about substantial decrease of zeta potential. Also the efficiency of the counter complex ion to decrease the electrokinetic potential depends on its size and charge. The situation is also depicted in the ζ – log C curves of figure 6.

Effect of surfactants on the stability

The influence of some non-ionic and cationic surfactants on the stability of the hydroxide emulsions was examined through their influence on zeta potential.

The non-ionic detergents used of Triton *X* series which are alkyl phenyl ethers of polyethylene glycol having the formulas:

Triton *X*-114 $= C_8H_{17}C_6H_5(CH_2CH_2O)_{7.5}$,

Triton *X*-100 $= C_8H_{17}C_6H_5(CH_2CH_2O)_{9.7}$,

Triton *X*-102 $= C_8H_{17}C_6H_5(CH_2CH_2O)_{12.3}$,

Triton *N*-101 $= C_9H_{19}C_6H_5(CH_2CH_2O)_{9.7}$,

the latter being isooctyl methyl phenyl derivative. It can be seen from the table 6 that these detergents get adsorbed at the hydroxide coated oil-water interface and thereby lower the zeta potential of the emulsion droplets. This is also reflected by the zeta-log C curves of figure 7 which have the same trend as curves of figure 6. It may be noted that although these detergents are able to decrease the zeta potential but unable to affect the charge reversal. Their efficiency is in the order Triton *X*-102 > Triton *N*-101 > Triton *X*-100 > Triton *X*-114 which is consistent with their

Table 6. Effect of addition of non-ionic detergents on $Th(OH)_3$ stabilized emulsions

Detergent concentration	Zeta potential in mV			
	Triton X-102	Triton N-101	Triton X-100	Triton X-114
1.0×10^{-7} M	31	33	35	36
5.0×10^{-7} M	28	30	33	35
1.0×10^{-6} M	23	26	29	31
5.0×10^{-6} M	16	20	22	25
1.0×10^{-5} M	12	18	20	22
5.0×10^{-5} M	8	12	15	18

Table 7. Effect of addition of cationic detergents on solid stabilized emulsions

Solid emulgent	Charge reversal concentration in moles per litre			
	CTAB	*CPB*	*CPC*	*LPC*
$Al(OH)_3$	0.8×10^{-5}	1.5×10^{-5}	2.0×10^{-5}	3.5×10^{-4}
$Cr(OH)_3$	1.2×10^{-5}	2.2×10^{-5}	3.8×10^{-5}	4.7×10^{-4}
$Fe(OH)_3$	1.1×10^{-5}	1.8×10^{-5}	3.3×10^{-5}	4.1×10^{-4}
$Co(OH)_2$	0.5×10^{-5}	1.3×10^{-5}	1.4×10^{-5}	2.9×10^{-4}
$Ce(OH)_3$	1.6×10^{-5}	3.1×10^{-5}	4.9×10^{-5}	5.2×10^{-4}
$Th(OH)_4$	1.4×10^{-5}	2.6×10^{-5}	4.2×10^{-5}	–
$Ti(OH)_3$	0.4×10^{-5}	0.6×10^{-5}	0.9×10^{-5}	2.2×10^{-4}

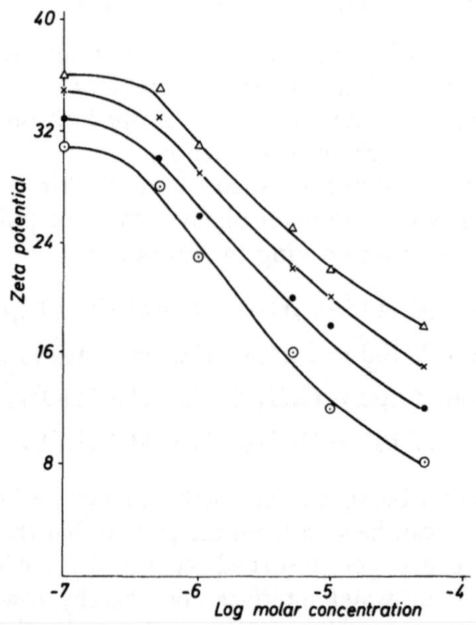

Fig. 7. Influence of non-ionic surfactants on zeta potential. ⊙ Triton *X*-102, ● Triton *N*-101, × Triton *X*-100, △ Triton *X*-114

chain length. These were, however, not as effective as the cationic detergents as discussed below. Moreover, below their CMC they have little effect which means that it is micellar adsorption (24) which effects the decrease of the zeta potential.

The cationic detergents were found to be more efficient in decreasing the zeta potential of the hydroxide stabilized emulsions. They are effective even below their CMC. Their continuous addition decreases the zeta more rapidly and ultimately reverses the charge. The charge reversal concentrations are listed in table 7 and indicate that their efficiency order (25) is CTAB > CPB > CPC > LPC which is also in accordance with their chain length. This is illustrated also diagrammatically in the charge reversal spectra of figure 8, which support the above mentioned observations. The curves also expose the behaviour of CPB and CPC which differ in the head group, the former being more efficient. Again of the seven emulsions examined $Ce(OH)_3$ stabilized seems to be the most stable whereas $Ti(OH)_2$ stabilized one is the least stable. This is in keeping with the previous observations. Lastly it may be noted that the charge reversal concentrations are higher than the corresponding CMC which implies that in charge reversal micellar adsorption plays the role.

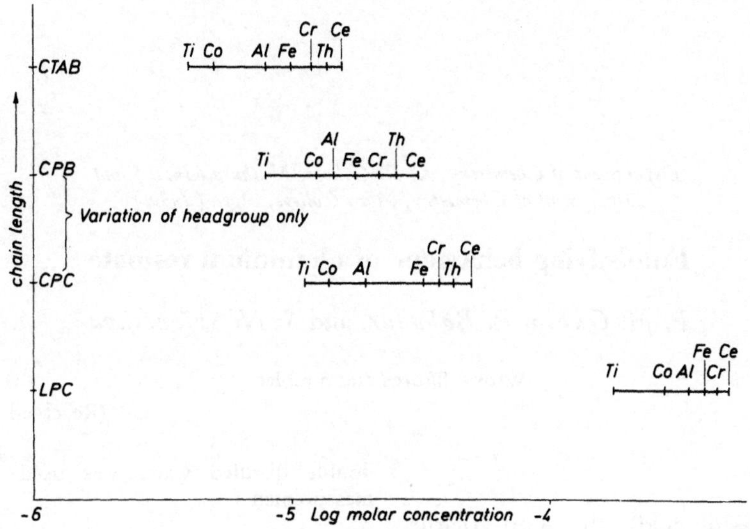

Fig. 8. Charge reversal spectra. The metal cation denotes the corresponding hydroxide as emulgent

Summary

Finely dispersed solids specially freshly precipitated hydroxides of polyvalent metals have got the property of promoting oil-in-water emulsions. A study of the hydroxides of 11 metals as emulsifiers reveals that they yield coarse, though very stable emulsions comparable to those stabilized by macromolecules. The emulsion behaviour has been interpreted by studying flocculation, coalescence and the electrokinetic potential. The latter has been studied as a function of electrolyte, complex salts and surfactants both cationic and non-ionic. The stability of the emulsions is attributed to the strength of the interfacial film of the adsorbed hydroxides which is presumably of colloidal dimensions.

Zusammenfassung

Die Arbeit behandelt die Wirkung feinverteilter, frisch gefällter Hydroxide mehrwertiger Metalle als Emulgatoren. Diese Hydroxide liefern relativ grobe Emulsionen. Ihre Stabilität ist vergleichbar mit entsprechenden Emulsionen, die durch Makromoleküle stabilisiert wurden. Die Stabilität wird auf die Stärke der Interfacial-Filme der adsorbierten Hydroxide zurückgeführt.

References

1) *Van der Minne, J. L.*, Over Emulsions, Amsterdam 66 (1928).

2) *Schulman, J. H., L. Leja*, Trans. Faraday. Soc. **50**, 598 (1954).

3) *Albers, W., J. Th. G. Overbeek*, J. Coll. Sci. **14**, 509 (1959).

4) *King, A., L. N. Mukerjee*, J. Soc., Chem. Ind. **58**, 243 (1939).

5) *Lucassen-Reynders, E. H.*, Stabilization of water-in-oil emulsions (Utrecht 1962).

6) *Vold, M. J., D. V. Rathnama*, J. Phys. Chem. **64**, 1619 (1960).

7) *Mukerjee, L. N., S. N. Srivastava*, Kolloid-Z. **147**, 146 (1956).

8) *Mukerjee, L. N., S. N. Srivastava*, ibid **149**, 35 (1956).

9) *Mukerjee, L. N., S. N. Srivastava*, ibid **150**, 144 (1957).

10) *Mukerjee, L. N., S. N. Srivastava*, ibid **150**, 148 (1957).

11) *Mukerjee, L. N., S. N. Srivastava*, ibid **162**, 10 (1959).

12) *Mukerjee, L. N., S. N. Srivastava*, ibid **170**, 32 (1960).

13) *Jain, S. P., S. N. Srivastava*, Bull. Chem. Soc. Japan **42**, 1171 (1969).

14) *Smoluchowski, M. V.*, Zeit. Physik Chem., **92**, 129 (1917).

15) *Verwey, R. J. W., J. Th. G. Overbeek*, Theory of Stability of Lyophobic Colloids (Amsterdam 1948).

16) *Derjaguin, B.*, Trans. Faraday Soc. **36**, 203 (1940).

17) *Eilers, H., J. Korff*, ibid **36**, 229 (1940).

18) *Derjaguin, B.*, Disc., Faraday Soc. **18**, 95 (1954).

19) *Srivastava, S. N., D. A. Haydon*, Proc. 4th Int. Cong.

20) *Srivastava, S. N.*, American Chemical Society Symposium Series Number 9, Colloidal Dispersions and Micellar Behaviour, 110 (1975).

21) *Srivastava, S. N.*, Koll. Zeit. **174**, 36 (1961).

22) *Van den Tempel, M.*, Stability of oil-in-water emulsions (Delf, Holland 1953).

23) *Matijevic, E., N. Kolak*, J. Colloid Interface Sci. **24**, 441 (1967).

24) *Bahadur, P., S. N. Srivastava*, Ind. J. Tech. **12**, 489 (1974).

25) *Sastry, T. G., S. N. Srivastava*, J. Colloid Interface Sci. **33**, 468 (1970).

Author's address:

S. N. Srivastava
UNESCO Science Centre
P.O. Box 832
Alexandria (Egypt)

Progr. Colloid & Polymer Sci. **63**, 50–54 (1978)

ISSN 0340-255 X

Department of Chemistry, K. R. College, Mathura (India) and
Department of Chemistry, Agra College, Agra (India)

Emulsifying behaviour of aluminium resinate

P. M. Gupta, P. Bahadur, and *S. N. Srivastava*

With 4 figures and 3 tables

(Received March 29, 1976)

Introduction

Rosin (or resin acid) the non-volatile fraction obtained by steam distillation of resin consists of a mixture of some isomeric diterpene acids, the most abundant being the abietic acid. Resin acid and its salts are mostly used as driers in varnishes, paints, etc. because of their surface active behaviour.

Aluminium soaps have a wide variety of technical uses (1, 2) and are thus very important compounds from that point of view. In addition to this they have also a great scientific interest since the colloidal properties of their solutions and their surface chemical behaviour give rise to many complicated structural problems. *Friberg* (3) has studied the relationship between the structure of various aluminium soaps and their colloid and surface behaviour in non-aqueous media.

Becher (4) in a monograph has reviewed the chemistry of emulsifying behaviour of soaps. Despite the work on the emulsifying properties of fatty acid soaps the studies with resin soaps are very scarce (5), which might be due to the fact that these soaps have got poor cleansing action.

However, the commercial use of resin soaps in paint industries is not less pronounced. Present studies are, therefore, made in order to examine the surface activity and emulsifying behaviour of such soaps.

Experimental

Materials

Aluminium resinate, the emulsifier was a BDH chemical. The inorganic salts used as flocculating reagents were BDH or E. Merck chemicals of AR grade. AnalaR Toluene was used as the oil phase and double distilled water was used throughout the measurements.

Procedure

All the stability measurements were carried out at 25° ± 1 °C in a constant temperature room. Changes in particle number on ageing and by the addition of flocculating salt were measured microscopically. A sample of emulsion was placed on to a hemocytometer of depth 0.1 mm and examined under a sensitive microscope under the magnification (40 × objective, 15 × eyepiece). The hemocytometer was thoroughly cleaned with chromic acid and washed with double distilled water and dried in a current of warm air. Traces of dust particles and grease were removed by wiping the surface with a tissue moistened with distilled acetone. About 300 particles were counted in a calibrated region. Duplicate counting was always carried out with a fresh slide and the mean was taken. The results were reproducible to 5%. Emulsions were prepared by shaking the mixture of 0.01 M KCl solution and toluene containing the emulsifier, such that the emulsifier concentration was 0.25% and the phase volume ration O/W being 1:19. The contents were shaken by hand for about ten minutes and were homogenized twice through a hand operated homogenizer which finally yielded emulsions with droplets of radius about one micron. These stock emulsions were diluted ten-folds by the 0.01 M KCl solution. KCl was added to maintain a constant ionic strength and to shorten the calculations of Debye-Hückel parameter and the zeta potential.

The electrophoretic mobility was measured in a rectangular micro-cell which was mounted laterally and the microscope was fixed at a stationary level. The emulsion globules were timed for two divisions of a calibrated graticule in alternate directions under the applied voltage. The calculations of zeta potential were made using Helmholtz equation. The mobility results were reproducible to 2%.

Results and discussion

The emulsion had the zeta potential 65.2 mV initially which decreased gradually with the

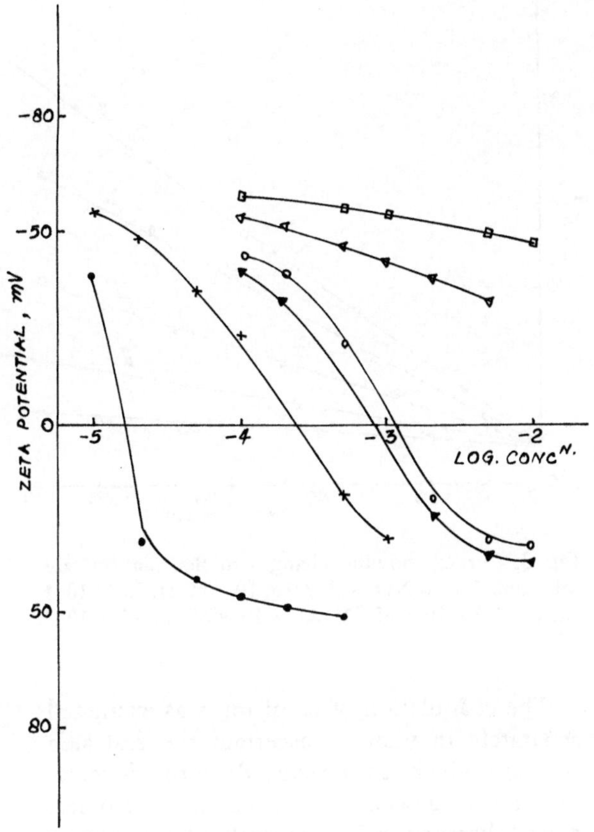

Fig. 1. Zeta potential – log C curves for ● $Th(NO_3)_4$, × $Cr(NO_3)_3$, ▲ $UO_2(NO_3)_2$, ○ $Pb(NO_3)_2$, △ $Ba(NO_3)_2$, □ $NaNO_3$

addition of inorganic salts, the decrease was, however, very small for Na^{+1} and Ba^{+2} ions. For bivalent UO_2^{+2} and Pb^{+2} ions the charge on the systems reduced to zero at about one millimolar concentration. Tri and tetravalent ions were very much effective in flocculating the emulsions, the flocculating powers of cations were in accordance of Hardy-Schulze rule (fig. 1).

After the charge reversal, an increase in zeta potential was observed on further addition of inorganic salts. After the complete flocculation of the system, which was seen at isoelectric point, the change in zeta potential was very sharp which was followed by constancy. The flocculation concentrations of the salts as determined at zero point of charge were 1.05×10^{-3} M, 7.95×10^{-4} M, 2.37×10^{-4} M and 1.68×10^{-5} M for $Pb(NO_3)_2$, $UO_2(NO_3)_2$, $Cr(NO_3)_3$ and $Th(NO_3)_4$.

Also to test the nature of flocculation and

Table 1. Energy data for aluminium resinate stabilized toluene-in-water emulsion flocculated by electrolytes

Electrolyte Concentration in mole/lit	Zeta potential mV	Height of maxima in kT	Distance at $V=0$ Å	Depth of the secondary minima in kT
$Th(NO_3)_4$				
1.0×10^{-5}	38.50	410.3	120	14.7
Charge reversed				
2.0×10^{-5}	32.29	211.6	110	18.9
5.0×10^{-5}	40.64	476.2	125	14.6
1.0×10^{-4}	45.00	639.8	130	13.6
2.0×10^{-4}	49.29	846.8	115	15.8
5.0×10^{-4}	50.08	849.8	95	20.00
$Cr(NO_3)_3$				
1.0×10^{-5}	55.89	1286.8	150	10.0
2.0×10^{-5}	48.15	931.8	140	10.5
5.0×10^{-5}	34.69	282.5	125	17.5
1.0×10^{-4}	22.75	21.10	85	22.8
Charge reversal				
5.0×10^{-4}	19.18	Flocculation in Primary minima		
1.0×10^{-3}	31.50	135.9	80	24.4
$UO_2(NO_3)_2$				
1.0×10^{-4}	39.69	447.2	115	14.7
2.0×10^{-4}	32.15	203.0	110	19.3
Charge reversal				
2.0×10^{-3}	25.43	13.6	75	25.2
5.0×10^{-3}	34.55	138.5	55	26.4
1.0×10^{-2}	36.91	101.6	25	31.1
$Pb(NO_3)_2$				
1.0×10^{-4}	43.01	566.8	105	12.7
2.0×10^{-4}	38.90	411.3	100	16.00
5.0×10^{-4}	20.00	Flocculation in primary minima		
Charge reversal				
2.0×10^{-3}	21.62	Flocculation in primary minima		
5.0×10^{-3}	31.15	66.7	55	24.4
1.0×10^{-2}	32.15	16.7	20	33.5
$Ba(NO_3)_2$				
1.0×10^{-4}	54.73	1200.8	145	10.0
2.0×10^{-4}	52.31	1044.8	120	13.7
5.0×10^{-4}	46.05	684.8	110	17.2
1.0×10^{-3}	41.00	443.1	105	22.7
2.0×10^{-3}	38.18	310.9	100	23.6
5.0×10^{-3}	32.00	84.0	55	26.4
$NaNO_3$				
1.0×10^{-4}	59.01	1482.8	150	10.0
5.0×10^{-4}	56.12	1289.8	140	10.5
1.0×10^{-3}	54.98	1199.8	130	11.0
5.0×10^{-3}	50.00	844.8	110	20.4
1.0×10^{-2}	47.29	644.8	100	24.3
5.0×10^{-2}	37.68	47.4	10	44.6

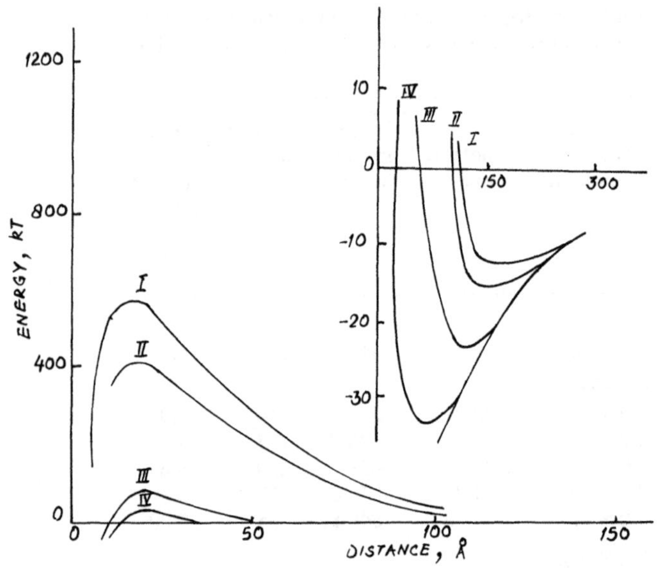

Fig. 2. The interaction energy diagram for $Pb(NO_3)_2$. I. 1.0×10^{-4} M, II. 2.0×10^{-4} M, III. 5.0×10^{-3} M, IV. 1.0×10^{-2} M

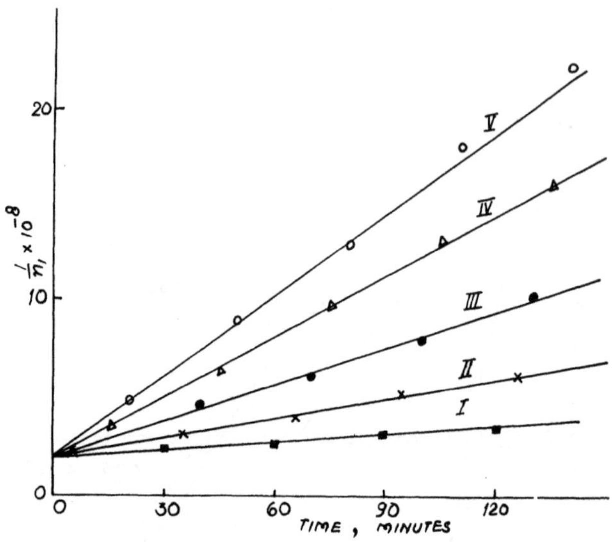

Fig. 3. Curves showing changes in flocculation rate with time for $Pb(NO_3)_2$. I. 2.0×10^{-4} M, II. 5.0×10^{-4} M, III. 1.0×10^{-3} M, IV. 2.0×10^{-3} M, V. 5.0×10^{-2} M

the applicability of DLVO theory (6) in the present systems the interaction energies were calculated by summing up the attraction and repulsion energies separately. The usual equation due to *Schenkel* and *Kitchener* (7) and *Derjaguin* and *Kussakov* (8) were used for the calculation of attraction and repulsive energies. The energy data are reported in table 1, and an energy diagram for $Pb(NO_3)_2$ shown in figure 2. It is obvious that high energy humps sometimes exceeding even thousands of kT were observed when the emulsion droplets were very close to one another. The emulsions, therefore, seemed to be stable against flocculation as it was improbable for emulsion droplets to aggregate by crossing such high energy barriers. Microscopic examination by observing the changes in particle concentration with time revealed that the emulsion showed appreciable flocculation with electrolyte concentration and on ageing. It was, therefore, assumed justified that the flocculation of the emulsions only occurs in secondary minima where the emulsion droplets were at larger distances roughly about 75–150 Å. This reversible flocculation of emulsion was further verified by redispersing the coagulated emulsions by shaking or by other mechanical means. However, in a few cases where the zeta potential was less than about 25 mV, the primary flocculation was also observed.

The coagulation of emulsion was examined separately in view to ascertain the real step through which it proceeds. Both the flocculation and coalescence of the emulsion droplets were observed microscopically by counting the total number of droplets present in a specified volume, whether they are associated or not.

The flocculation rate constant k_1, was determined from the slope of the line plotted between reciprocal number of primary particles against time, (fig. 3) whereas the coalescence rate constant was determined from the slope of the line plotted between logarithm of individual particles and time (fig. 4). Both these constants have been reported in table 2 for $Pb(NO_3)_2$ only as a representative for all other flocculating salts.

The increase in flocculation rate with electrolyte concentration was due to the partial neutralization of the surface charge. The values of coalescence rate constant were of the order of 10^{-3} sec^{-1} which showed that out of every one thousand oil droplets contained in aggregates two will coalesce per second.

The real step through which the coagulation of the system proceeds may be examined by calculating the ratio $\frac{k_1 n_0}{k_2}$. It appears from the ratio which was in all cases greater than unity

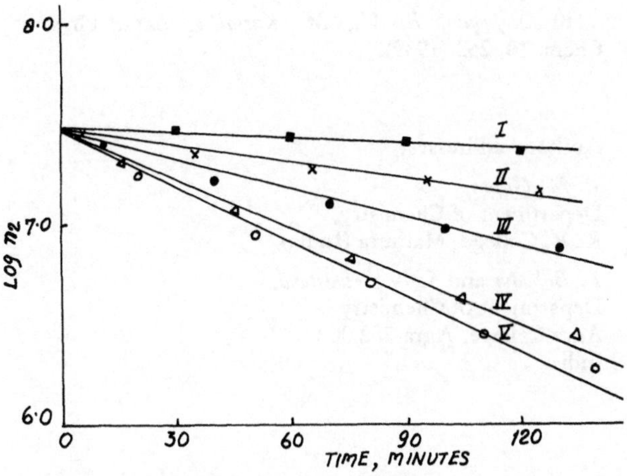

Fig. 4. Curves showing changes in coalescence rate with time for $Pb(NO_3)_2$. I. 2.0×10^{-4} M, II. 5.0×10^{-4} M, III. 1.0×10^{-3} M, IV. 2.0×10^{-3} M, V. 5.0×10^{-3} M

Table 2. Effect of $Pb(NO_3)_2$ on the coagulation rate constants

Electrolyte concentration mole/lit.	$k_1 \times 10^{10}$ cm^3 sec^{-1}	$k_2 \times 10^3$ sec^{-1}	$\frac{k_1 \times n_0}{k_2}$
2.0×10^{-4}	1.33	0.83	5.60
5.0×10^{-4}	4.00	3.00	4.66
1.0×10^{-3}	6.67	5.33	4.38
2.0×10^{-3}	10.67	8.67	4.31
5.0×10^{-3}	14.33	10.33	4.90

that the coalescence was slower and the rate determining step.

Binding parameters

From the gradients of ζ log c curves (fig. 1) at zero point of charge, the binding parameters like adsorption constants k_1 and k_2, the number of binding sites N_1, the binding energy ΔG and the charge density in the Stern layer were calculated. The values are reported in table 3. The adsorbabilities of cations, as denoted by these values seem in the order as found previously. The ΔG values were of the order of 6–10 K Cals/mole which seem to be reasonably high as expected on the basis of electrostatic adsorption on potential determining sites. Since the ΔG values were calculated at isoelectric point, these must be very close to electrochemical free energy.

Summary

Stable oil-in-water emulsions were prepared by the use of aluminium resinate as emulgent. The stability was determined in terms of coagulation which was brought about by inorganic salts with different cations. Coagulation of the emulsion was studied hemocytometrically and electrophoretically. Kinetically coalescence was found to be the rate determining step. Electrophoretic data were recorded as a function of electrolyte concentration. The charge reversal was observed for Th^{4+}, Cr^{3+}, UO_2^{2+} and Pb^{2+} ions. Using particle size, zeta potential and van der Waals constant, as parameters to calculate interaction energies, it was found that the DLVO theory allows satisfactory quantitative interpretation of emulsion stability. Various parameters for cationic binding were calculated at the isoelectric point of the emulsion.

Zusammenfassung

Es wurden stabile Öl-in-Wasser-Emulsionen unter Verwendung von Aluminiumresinat als Emulgator hergestellt und ihre Stabilität gegen anorganische Salze mit unterschiedlichen Kationen bestimmt. Die Koagulation der Emulsionen wurde haemocytometrisch und elektrophoretisch verfolgt. Es wurde gefunden, daß kinetisch die Koaleszenz der geschwindigkeitsbestimmende Schritt ist. Die elektrophoretischen Daten wurden als Funktion der Elektrolytkonzentration gemessen. Für Th^{4+}, Cr^{3+}, UO_2^{2+} und Pb^{2+} wurde Ladungsumkehr beobachtet. Die DLVO-Theorie erlaubt eine befriedigende quantitative Interpretation der Stabilität der Emulsionen.

Table 3. Parameters for cationic bindings

Cation	flocculation concentration moles/lit	$K_1 \times 10^{-8}$	$K_2 \times 10^{-4}$	$N_1 \times 10^{-12}$ groups/cm^2	$\sigma S \times 10^{-3}$ Coulombs	$-\Delta G$ Kcals/mole
Th^{+4}	1.68×10^{-5}	13.73	24.40	2.93	4.52	9.46
Cr^{+3}	2.37×10^{-4}	0.44	0.49	6.31	4.86	7.20
UO_2^{+2}	7.95×10^{-4}	0.11	0.10	11.56	4.96	6.30
Pb^{+2}	1.05×10^{-3}	0.09	0.08	11.92	5.14	6.14

References

1) *Pilpel, N.*, Res. Appl. Ind. **15**, 385 (1962).
2) *Friberg, S.*, Surface Chemistry, 203 (Copenhagen 1965).
3) *Friberg, S.*, Svensk Kemisk Tidskrift **78**, 1 (1966).
4) *Becher, P.*, Emulsions-Theory and Practice, p. 209 (New York 1965).
5) *Markley, K.*, Fatty acid (1961), Vol. II, p. 715 (New York 1961).
6) *Verwey, E. J. M., J. Th. G. Overbeek*, Theory of Stability of Lyophobic Colloids (Amsterdam 1948).
7) *Schenkel, J. N., J. A. Kitchener*, Trans. Faraday Soc. **56**, 161 (1960).
8) *Derjaguin, B. V., M. Kussokov*, Acta. Phys. Chem. **10**, 253 (1939).

Authors' addresses:

P. M. Gupta,
Department of Chemistry,
K. R. College, Mathura (India)

P. Bahadur and *S. N. Srivastava*,
Department of Chemistry,
Agra College, Agra-282 002
India

Progr. Colloid & Polymer Sci. **63**, 55–64 (1978)

ISSN 0340-255 X

Chemical Laboratories, Agra College, Agra (India)

Effect of some cationic surfactants on an oil/water emulsion stabilized by dodecyl benzene sulphonate

M. K. Sharma, G. Chandra, S. K. Jha, and *S. N. Srivastava*

With 8 figures and 4 tables

(Received March 7, 1977)

Introduction

The surfactants are of considerable importance in emulsion studies due to their surface activity. Recently, the effect of some anionic surfactant on the stability of chromium hydroxide stabilized emulsion has been studied by *Jain* and *Srivastava* (1, 2). *Haydon* (3) has studied the relation between electrokinetic potential and surface charge density of some surfactants stabilized oil-in-water emulsions. Recently, *Sharma* and *Srivastava* (4, 5) have studied the effect of some ionic surfactants on the emulsions stabilized by some drugs. The interaction of some surfactants on the interfacial activity of sodium dodecyl sulphate has also been reported by *Bahadur* et al. (6, 7). Kinetics of coagulation of stable emulsions have been investigated by *Miyamato* (8). The present work deals with the role of some cationic surfactants on the stability of dodecyl benzene sulphonate stabilized oil-in-water emulsions in the light of modern electrical theories. The adsorption constants have also been calculated to throw some light on the nature of adsorption.

Materials and methods

Lauryl pyridinium chloride (LPyCl) was supplied by Industrial Chemical Products, Bombay, India. The product was extracted with acetone and then recrystallization process was repeated till finally white long needles were obtained. All other cationic surfactants i.e. tetra decyl pyridinium bromide (TPyBr), cetyl trimethyl ammonium bromide (CTABr), cetyl pyridinium chloride (CPyCl) and cetyl pyridinium bromide (CPyBr) were obtained from British Drug Houses Ltd., Poole, England, and used without further purification. The other chemicals and reagents used were analar grade obtained from BDH.

The emulsions were prepared by dispersing 4% by volume of the xylene in aqueous solution of 0.1% dodecyl benzene sulphonate. The final mixture also contained 0.01 M KCl. The mixture was hand shaken first about half an hour and then passed twice through a stainless steel homogenizer. In all estimation emulsions were prepared and sampled under identical conditions so as to have maximum precision.

The flocculation of the globules was observed by counting the number of particles with respect to time using double Neubauer hemocytometer under the Olympus microscope. For the size of the globules the microphotograph was taken using a magnification 15×40 times, which resulted in average particle diameter 1.02 μ. The electrokinetic potentials were determined in the presence of cationic surfactants by the Northrop-Kunitz (9) type rectangular cell microelectrophoretically using the Helmholtz-Smoluchowski (10) equation.

Calculations

The charge density of emulsion globules in the Stern layer in presence of cationic surfactants is given by the Stern equation.

$$\sigma_s = \frac{N_1 ev}{1 + \frac{55 \cdot 6}{C} \exp\left(\frac{-\Delta G}{kT}\right)} \qquad [1]$$

where C is the molar concentration of the surfactants, N_1 is the number of binding sites available per cm² and ΔG is the free energy of adsorption per molecule. This can also be written in the form:

$$\sigma_s = \frac{k_1 C}{1 + k_2 C}$$

where

$$k_1 = N_1 k_{2ev}$$
$$k_2 = \exp(-\Delta G/kT)/55 \cdot 6. \quad [2]$$

The adsorption constants k_1 and k_2 have been calculated using following treatment given by *Ottewill* et al. (11) at zero point of charge

$$\frac{1}{C} = \frac{4\pi e v N_1 k_2}{\varkappa \varepsilon \psi_0} - k_2 \quad [3]$$

$$\left(\frac{d\psi_{0\delta}}{d \ln C}\right)_{\psi_\delta=0} = \left(\frac{\varkappa \varepsilon \psi_{0\delta}}{4\pi e v N_1} - 1\right) \psi_{0\delta} \, . \quad [4]$$

The value of the factor $\left(\frac{d\psi_{0\delta}}{d \ln C}\right)_{\psi_{0\delta}=0}$ is taken from the slope of the log c-electrokinetic potential curves. C is here the concentration of the surfactants at isoelectric point.

The coagulation of an emulsion comprises flocculation and coalescence. Kinetically, the flocculation is bimolecular reaction and is represented by the equation:

$$-\frac{dn_1}{dt} = K_1 n_1^2$$

or

$$1/n_1 - 1/n_0 = K_1 t \quad [5]$$

where n_0 is the number of globules present initially and n_1 is the number of primary particles taking part in the flocculation process.

Kinetically, the coalescence is unimolecular reaction

$$-\frac{dn_2}{dt} = K_2 n_2$$

$$n_2 = n_0 e^{-K_2 t} \quad [6]$$

where n_2 is the number of individual emulsion globules at any time $t \cdot K_1$ and K_2 are the rate constants for flocculation and coalescence respectively.

The relative values of flocculation and coalescence rates are evaluated by the ratio $K_1 n_0/K_2$ which is less than unity if the coalescence is rapid and the overall rate determined entirely by flocculation and vice versa.

For the discussion of the stability of the emulsion system, the total interaction energy (V) has been calculated by the following equations (12–14).

$$V = V_R + V_A \quad [7]$$

where V_R and V_A are the repulsive and attractive energies respectively. V_R is given by the following equation (15).

$$V_R = \frac{\varepsilon a \psi_0^2}{2} \ln (1 + e^{-\varkappa H}) \quad [8]$$

where ψ_0 is the surface potential and assumed to be equal to the electrokinetic potential. ε is the dielectric constant, H is the interparticle distance and $\varkappa$ is the Debye-Hückel parameter.

The attractive energy (V_A) is given by the following equations (16).

$$V_A = -\frac{Aa}{12\,H}\left(\frac{\lambda}{\lambda + 3 \cdot 54\, \pi H}\right) \quad \text{for } H < 150 \text{ Å} \quad [9]$$

$$V_A = -\frac{Aa}{\pi}\left(\frac{2 \cdot 45\, \lambda}{120\, H^2} - \frac{\lambda^2}{1045\, H^3} + \frac{\lambda^3}{5 \cdot 62 \times 10^4\, H^4}\right) \quad \text{for } H > 150 \text{ Å} \quad [10]$$

where A is the van der Waals constant. λ is the wavelength of the intrinsic electronic oscillations of the atoms (approximately equal to 10^{-5} cm) and a is the mean radius of the globules.

The theoretical value of degree of aggregation is calculated as a function of A from the formula

$$D = 4\pi a^3 n_0 \int S^2 \exp(-V/kT) ds \quad [11]$$

where $S = 2 + H/a$.

The integrant of this equation is evaluated graphically.

The degree of aggregation can also be obtained practically by counting the monomers and dimers hemocytometrically by the following relation:

$$D = \frac{\text{Number of dimers}}{\text{Number of monomers} + \text{Number of dimers}} \quad [12]$$

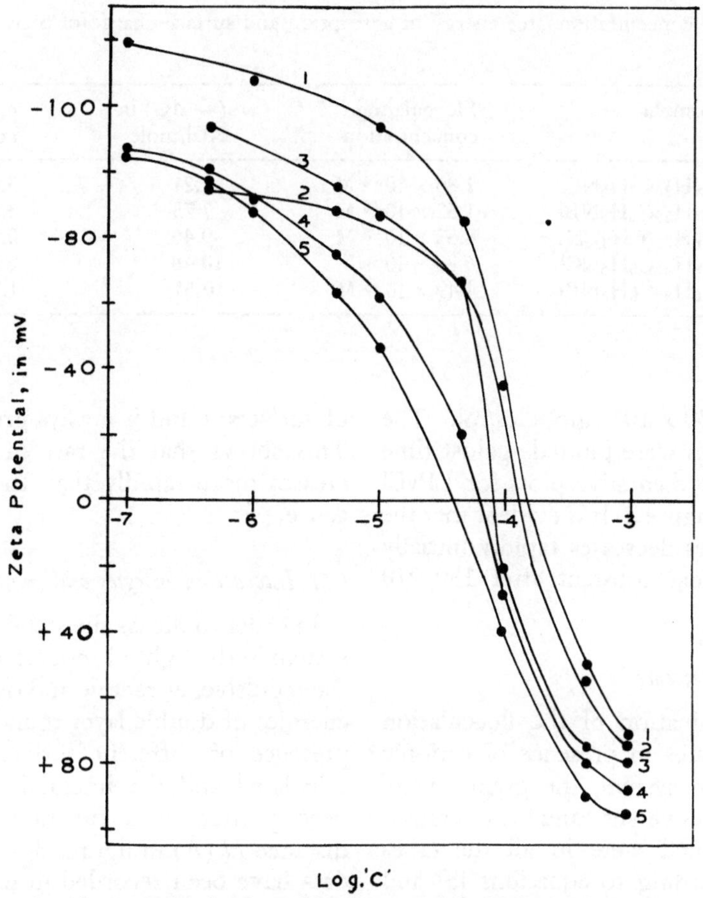

Fig. 1. Curves showing the variation in electrokinetic potential with log molar conc. of surfactants. (1) LPyCl (2) TPyBr, (3) CTABr, (4) CPyCl, (5) CPyBr

Results and discussion

(a) The effect of cationic surfactants on electrokinetic potential of the system

The value of the electrokinetic potential for the system in absence of surfactant was found to be 145.2 mV. The electrokinetic potentials in presence of different surfactants have been calculated and plotted against log molar concentrations as given in figure 1. All the surfactants reversed the charge of the system and their concentrations at zero point of charge were 1.80×10^{-4} M, 1.62×10^{-4} M, 8.53×10^{-5} M, 6.89×10^{-5} M and 4.49×10^{-5} M for LPyCl, TPyBr, CTABr, CPyCl and CPyBr respectively. Thus CPyBr is the most powerful surfactant for reversing the charge of the system.

While the surfactants with similar head group and different chain length i.e. LPyCl and CPyCl differ noticeably in their efficiency to reverse the charge (flocculating concentrations: 1.80×10^{-4} M for LPyCl and 6.89 $\times 10^{-5}$ M for CPyCl), the surfactants with different head groups and similar chain lengths i.e. CTABr, TPyBr and CPyBr show no remarkable difference in their flocculating concentrations (table 1). Thus, the surfactants with longer chain length have been found more effective. Almost similar results were noticed by *Sastry* and *Srivastava* (17) in case of emulsion stabilized by some biological materials and by *Sharma* et al. (18) in o/w emulsion stabilized by some drugs.

(b) The effect of cationic surfactants on flocculation

The flocculation studies were made in presence of different amount of surfactants by counting the number of unassociated globules present in a definite volume at various intervals of time. The initial number of globules was

Table 1. Flocculating concentration, free energy of adsorption and surface charge of Stern layer for different surfactants

Surfactants	Formula	Flocculating concentration	$(-\Delta G)$ in KCl/mole	σ_s μcb/mol.
LPyCl	$C_{12}H_{25}C_2H_5NCl$	1.80×10^{-4} M	7.21	3.86
TPyBr	$C_{14}H_{29}C_5H_5NBr$	1.62×10^{-4} M	7.75	3.80
CTABr	$C_{16}H_{33}(CH_3)_3NBr$	8.53×10^{-5} M	9.46	2.94
CPyCl	$C_{16}H_{33}C_5H_5NCl$	6.89×10^{-5} M	10.48	2.06
CPyBr	$C_{16}H_{33}C_5H_5NBr$	4.49×10^{-5} M	10.51	1.72

found to be 4.30×10^8 globules/ml. The number of globules were plotted against time and one of the representative plots for LPyCl has been given in figure 2. It is evident that the number of globules decreases rapidly initially and becomes almost constant after 180–210 minutes.

(c) The rate of coalescence

For the determination of the flocculation and coalescence rates in presence of cationic surfactants for the system, the graphs were plotted between $1/n_1$ versus t and ln n_2 versus t. The curves obtained were in all the cases straight lines according to equations [5] and [6]. It was found that both the flocculation and coalescence rates decrease with decreasing concentration of cationic surfactants. It has also been observed that the ratio of $K_1 n_0/K_2$ slightly increases with increasing concentration of surfactant and is always greater than unity. This shows that the rate of flocculation increases more rapidly than the rate of coalescence.

(d) Interaction energies and emulsion stability

In order to dicuss the stability of the present system in the light of modern electrical theory, the repulsive, attractive and overall interaction energies of double layer round the globules in presence of different surfactants have been calculated and the interaction energies have been plotted as a function of interparticle distance H (Å) in figures 3–7 and the relevant data have been recorded in table 2. The high energy maxima in most cases ranging between 1000–5000 kT shows that the emulsions are stable and flocculation cannot occur in the primary minima. Hence flocculation is only possible to occur in secondary minima.

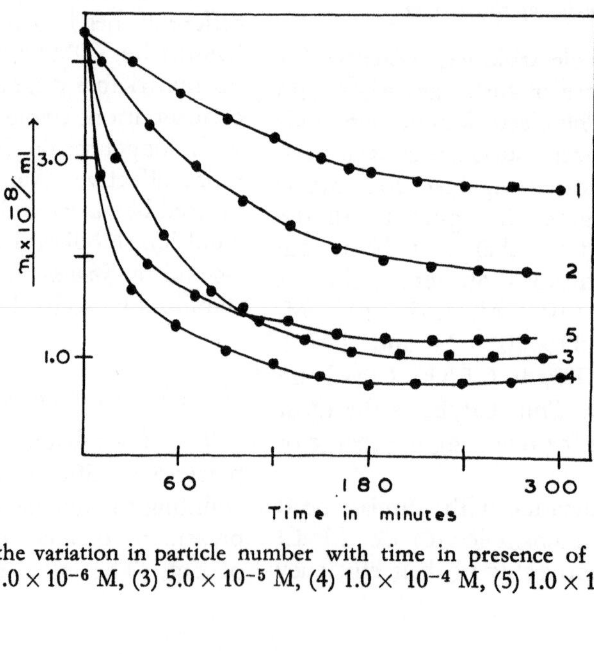

Fig. 2. Curves showing the variation in particle number with time in presence of LPyCl. (1) 1.0×10^{-7} M, (2) 1.0×10^{-6} M, (3) 5.0×10^{-5} M, (4) 1.0×10^{-4} M, (5) 1.0×10^{-3} M

(e) Secondary minimum flocculation

The high values of potential energy show the high degree of stability of the emulsion system. The flocculation on the basis of existence of secondary minimum flocculation can be explained. It is evident that at larger distance repulsion becomes negligible and the attraction will predominate. Therefore, the secondary minimum flocculation will occur.

The elaborated secondary minimum curves for different concentration of surfactants have been represented in figures 3–7. It is evident from these figures and table 2 that in case of low concentration range of all surfactants the depth of the secondary minimum is about 4.00 *kT* which is too shallow to trap the globules but at higher concentrations the minima has deepened upto 15–17 *kT* which are reasonably deep for reversible aggregation of the emulsion globules. The nature of flocculation has also been justified by plotting graphs between the number of monomers against time (fig. 2).

(f) Degree of aggregation (D) and van der Waals constant (A)

To evaluate the value of van der Waals constant the theoretical values of degree of aggregation for different values of van der Waals constants have been calculated with the help of equation [11]. These values of *D* have been plotted as a function of van der Waals constant in figure 8. The values of the degree of aggregation have also been estimated experimentally by counting the number of monomers and dimers, using equation [12]. The values of van der Waals constant (*A*) satisfying the observed values of degree of aggregation (*D*) are taken from figure 8 and

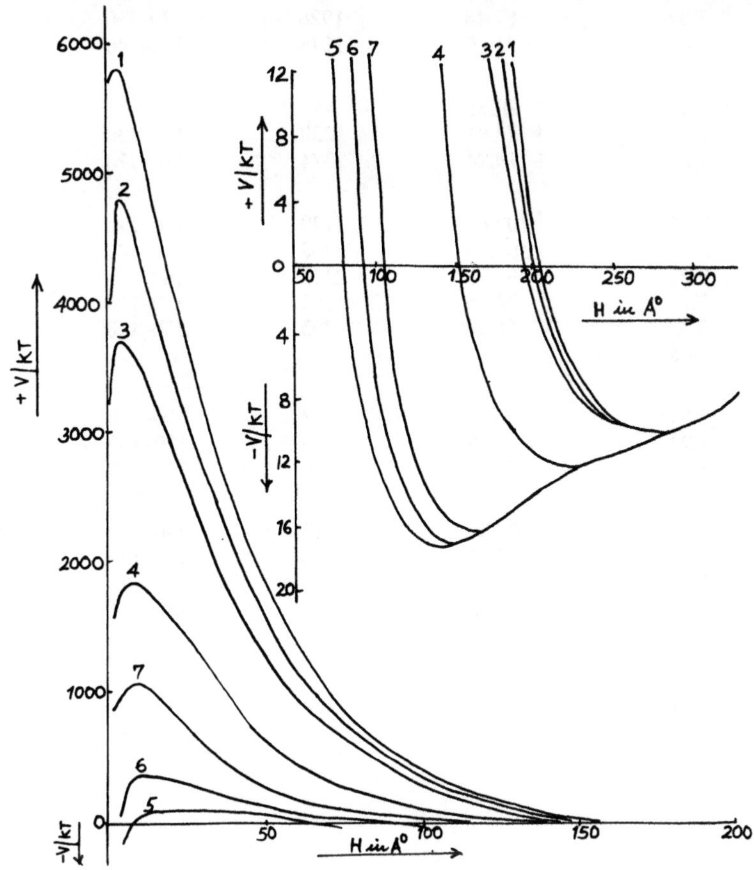

Fig. 3. Curves showing the variation in interaction energy as a function of interparticle distance in presence of LPyCl. (1) 1.0×10^{-7} M, (2) 1.0×10^{-6} M, (3) 1.0×10^{-5} M, (4) 5.0×10^{-5} M, (5) 1.0×10^{-4} M, (6) 5.0×10^{-4} M, (7) 1.0×10^{-3} M

Table 2. Electrokinetic potential, Debye-Hückel parameter, height of maxima, interparticle distance at which energy equal to zero and depth of the secondary minima with different concentration of surfactants

Surfactant concentration	Debye-Hückel parameter $\varkappa \times 10^{-6}$	Electrokinetic potential in mV	Height of maxima in kT	Interparticle distance at $V=0$ in Å	Depth of secondary minima in kT
LPyCl					
1.0×10^{-7} M	3.21	139.97	5809	200–210	9.8
1.0×10^{-6} M	3.25	128.30	4784	190–200	9.8
1.0×10^{-5} M	3.32	114.86	3721	190–200	10.0
5.0×10^{-5} M	3.45	86.51	1824	150–160	12.1
1.0×10^{-4} M	3.54	35.48	47	80– 90	17.4
Charge reversal					
5.0×10^{-4} M	3.94	+ 49.57	365	90–100	17.0
1.0×10^{-3} M	4.24	+ 71.41	1058	100–110	16.3
TPyBr					
5.0×10^{-7} M	3.24	95.09	2348	170–180	10.7
5.0×10^{-6} M	3.29	89.59	2033	170–180	10.7
5.0×10^{-5} M	3.45	65.12	900	130–140	13.2
Charge reversal					
5.0×10^{-4} M	3.94	+ 54.82	516	90–100	16.1
1.0×10^{-4} M	4.24	+ 74.21	1167	100–110	16.4
CTABr					
5.0×10^{-7} M	3.24	114.46	3499	180–190	10.2
5.0×10^{-6} M	3.29	96.52	2445	180–190	10.4
1.0×10^{-5} M	3.32	87.48	1928	170–180	10.7
5.0×10^{-5} M	3.45	71.44	1146	130–140	12.7
Charge reversal					
1.0×10^{-4} M	3.54	+ 20.83	–	–	–
5.0×10^{-4} M	3.94	+ 74.91	1230	120–130	14.6
1.0×10^{-3} M	4.24	+ 79.24	1374	110–120	15.8
CPyCl					
1.0×10^{-7} M	3.21	107.62	3179	180–190	10.8
5.0×10^{-7} M	3.24	101.53	2758	170–180	11.0
1.0×10^{-6} M	3.25	92.71	2195	160–170	11.4
5.0×10^{-6} M	3.29	74.56	1295	150–160	11.9
1.0×10^{-5} M	3.32	61.92	795	130–140	12.8
5.0×10^{-5} M	3.45	–	–	–	–
Charge reversal					
1.0×10^{-4} M	3.54	+ 28.74	27	50– 60	13.4
5.0×10^{-4} M	3.94	+ 79.96	1445	120–130	14.3
1.0×10^{-3} M	4.24	+ 86.02	1716	110–120	15.2
CPyBr					
1.0×10^{-7} M	3.21	105.30	3017	180–190	10.6
5.0×10^{-7} M	3.24	97.04	2418	170–180	10.9
1.0×10^{-6} M	3.25	87.16	1889	160–170	11.2
5.0×10^{-6} M	3.29	63.26	846	140–150	12.6
1.0×10^{-5} M	3.32	46.53	315	110–120	14.2
Charge reversal					
1.0×10^{-4} M	3.44	+ 39.84	140	80– 90	12.5
5.0×10^{-4} M	3.94	+ 90.02	2031	130–140	13.8

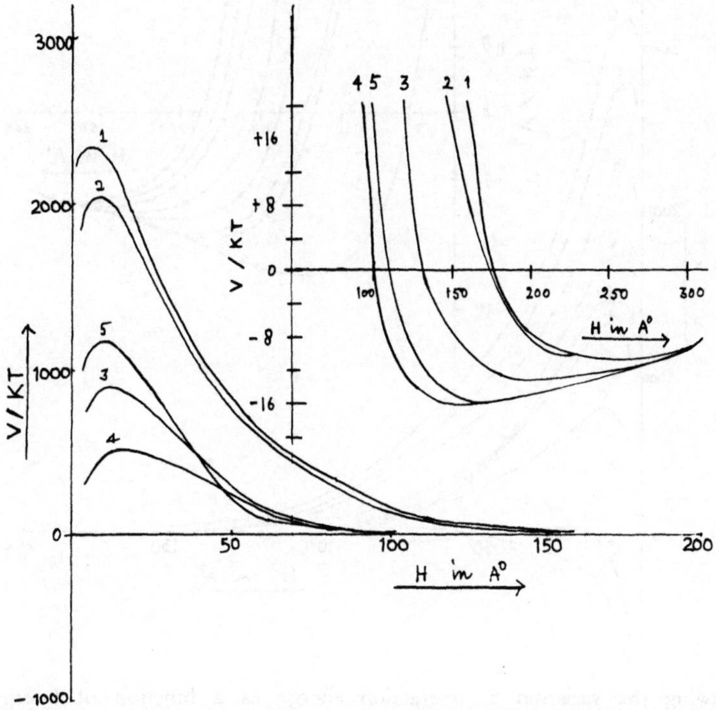

Fig. 4. Curves showing the variation in interaction energy as a function of interparticle distance in presence of TPyBr. (1) 5.0×10^{-7} M, (2) 5.0×10^{-6} M, (3) 5.0×10^{-5} M, (4) 5.0×10^{-4} M, (5) 1.0×10^{-3} M

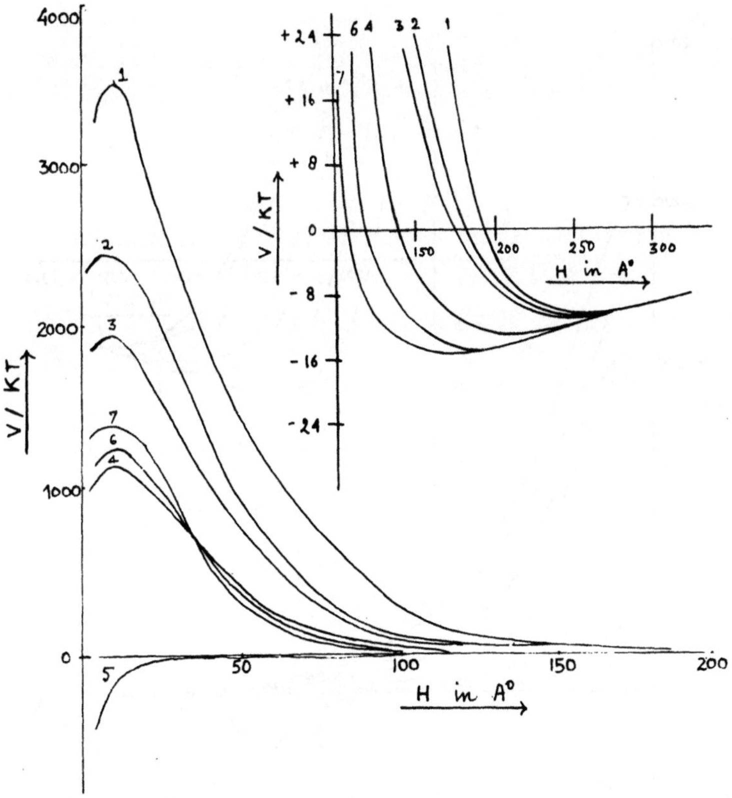

Fig. 5. Curves showing the variation in interaction energy as a function of interparticle distance in presence of CTABr. (1) 5.0×10^{-7} M, (2) 5.0×10^{-6} M, (3) 1.0×10^{-5} M, (4) 5.0×10^{-5} M, (5) 1.0×10^{-4} M, (6) 5.0×10^{-4} M, (7) 1.0×10^{-3} M

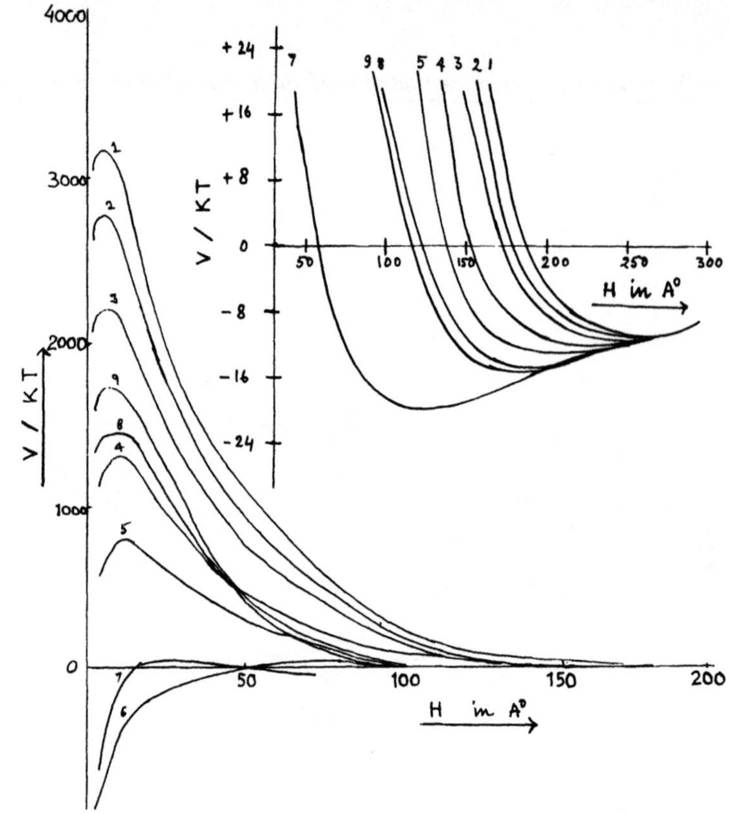

Fig. 6. Curves showing the variation in interaction energy as a function of interparticle distance in presence of CPyCl. (1) 1.0×10^{-7} M, (2) 5.0×10^{-7} M, (3) 1.0×10^{-6} M, (4) 5.0×10^{-6} M, (5) 1.0×10^{-5} M, (6) 5.0×10^{-5} M, (7) 1.0×10^{-4} M, (8) 5.0×10^{-4} M, (9) 1.0×10^{-3} M

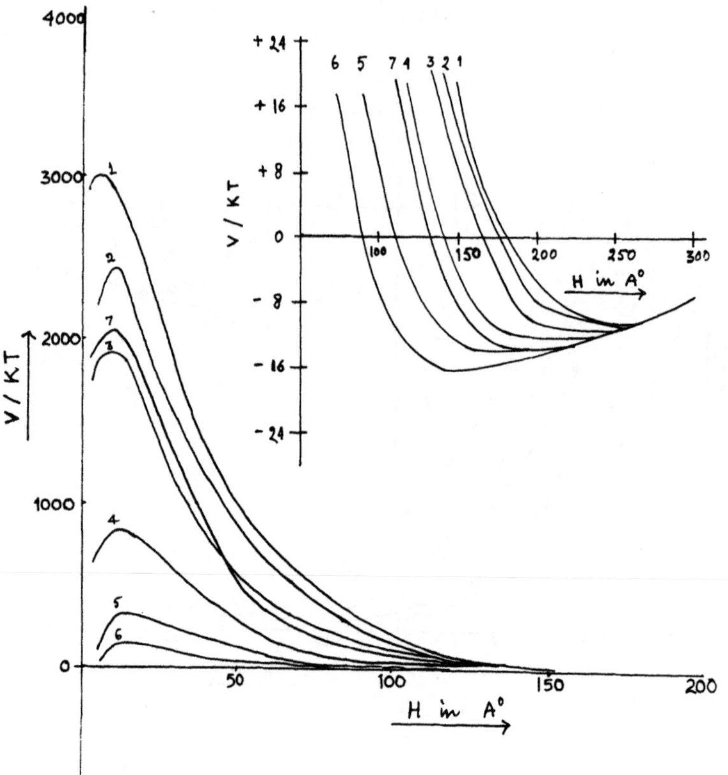

Fig. 7. Curves showing the variation in interaction energy as a function of interparticle distance in presence of CPyBr. (1) 1.0×10^{-7} M, (2) 5.0×10^{-7} M, (3) 1.0×10^{-6} M, (4) 5.0×10^{-6} M, (5) 1.0×10^{-5} M, (6) 1.0×10^{-4} M, (7) 5.0×10^{-4} M

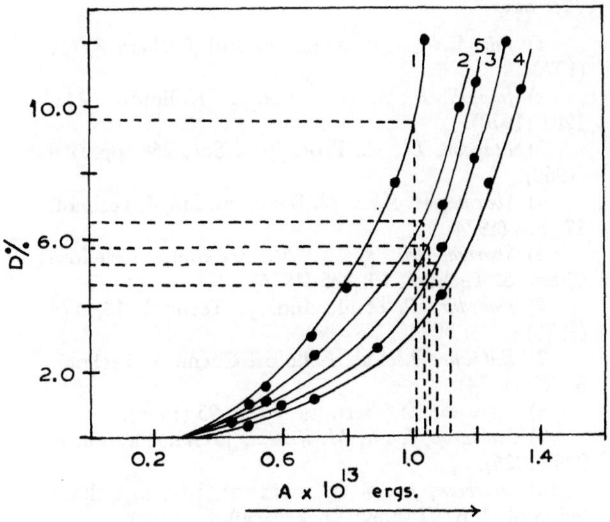

Fig. 8. Curves showing the variation in D with van der Waals constant (A). (1) 1.0×10^{-4} M, (2) 5.0×10^{-4} M, TPyBr, (3) 5.0×10^{-5} M, CTABr, (4) 1.0×10^{-6} M, CPyCl, (5) 1.0×10^{-6} M, CPyBr

recorded in table 3. The mean value of van der Waals constant was obtained 1.06×10^{-13} erg which agrees with the values obtained by earlier workers (18).

(g) Charge densities and adsorption constants

The adsorption constants k_1 and k_2, charge density in Stern layer σ_s, number of binding sites available N_1, and free energy of adsorption have been calculated with the help of flocculating concentrations using the equations given in the theoretical part and data obtained have been recorded in tables 1 and 4 for different surfactants. It is obvious from table 4 that k_2 remains always in the order of 10^5 for cetyl salts with different head groups. A remarkable change in k_2 values is noticed for LPyCl and CPyCl, surfactants with different chain lengths. Therefore, it is observed that the effect of chain length is more pronounced than the head groups.

Further, it is noticed that the ratio of k_1/k_2 values is almost in the same region for all the surfactants. Thus the agreement between the theory and experiment is good and shows that the coagulation with surfactants can be satisfactorily accounted by the influence of adsorption on surface potential. The values of the number of binding sites decrease with increasing chain length of surfactants. This is also confirmed by the data on free energy of adsorption, greater is the amount of counter ions to reach the critical value of charge reversal concentration. The values of free energy of adsorption are obtained in the range of 7.00 to 11.00 K.cal/mole, which are high enough for adsorption.

Table 3. Theoretical values of van der Waals constant obtained graphically by the observed degree of aggregation

Surfactants		Value of A corresponding to D observed (erg $\times 10^{13}$)	Mean value of A
1.0×10^{-4} M	LPyCl	1.01	
5.0×10^{-4} M	TPyBr	1.04	
5.0×10^{-5} M	CTABr	1.08	$= 1.06 \times 10^{-13}$ erg
1.0×10^{-6} M	CPyCl	1.12	
1.0×10^{-6} M	CPyBr	1.05	

Table 4. Adsorption constants and number of binding sites available per cm² for different surfactants

Surfactants	k_1	k_2	k_1/k_2	Binding sites N_1 groups/cm²
Chain length				
LPyCl	1.07×10^8	3.67×10^3	2.91×10^4	6.08×10^{13}
TPyBr	1.74×10^8	9.09×10^3	1.91×10^4	3.99×10^{13}
CPyBr	5.17×10^9	9.80×10^5	0.53×10^4	1.10×10^{13}
Head Group				
CTABr	1.57×10^9	1.66×10^5	0.95×10^4	1.98×10^{13}
CPyCl	5.87×10^9	9.34×10^5	0.63×10^4	1.31×10^{13}

Acknowledgement

Authors wish to express their thanks to the Principal, Agra College, Agra, for providing research facilities and also to the Council of Scientific and Industrial Research, New Delhi, for the award of a Post-Doctoral Fellowship to one of them (MKS).

Summary

The effect of some cationic surfactants on the stability of the xylene-in-water emulsion stabilized by dodecyl benzene sulphonate has been studied. The globules size was determined microphotographically. The stability of the emulsion system has been assessed by the measurement of electrokinetic potentials in presence of surfactants. The interaction energies have also been calculated and plotted as a function of interparticle distance. Flocculation studies were carried out hemocytometrically and values were correlated with electrokinetic potential. The flocculation and coalescence rate constants were determined separately in order to study the kinetics of coagulation of the system. The nature of adsorption of these surfactants has also been explained by calculating adsorption constants, value of van der Waals constant, number of binding sites, charge density and free energy of adsorption.

Zusammenfassung

Es wurde der Einfluß einiger kationischer Tenside (Alkylpyridiniumsalze) auf die Stabilität von Xylol-in-Wasser-Emulsionen, die durch Dodecylbenzolsulfonat stabilisiert waren, untersucht. Die Stabilität der Emulsionen wurde durch Messung der elektrokinetischen Potentiale in Gegenwart der Tenside abgeschätzt. Außerdem wurden die Wechselwirkungsenergien berechnet. Zur Verfolgung der Kinetik der Koagulation wurden die Geschwindigkeitskonstanten der Flockung und der Koaleszenz getrennt bestimmt. Darüber hinaus wurde die Adsorption der Tenside näher untersucht.

References

1) *Jain, S. P., S. N. Srivastava*, Ind. J. Chem. **8**, 731 (1970).

2) *Jain, S. P., S. N. Srivastava*, Kolloid-Z. **242**, 1201 (1970).

3) *Haydon, D. A.*, Proc. Roy. Soc. **258**, pp. 319 (1960).

4) *Sharma, M. K.; S. N. Srivastava*, Ind. J. Technol. **12**, 518 (1974).

5) *Sharma, M. K., S. N. Srivastava*, Cellulose Chem. & Technol, **10**, 235 (1976).

6) *Bahadur, P.* et al., Ind. J. Technol. **13**, 178 (1975).

7) *Bahadur, P.* et al., Cellulose Chem. & Technol. **8**, 387 (1974).

8) *Miyamato, S.*, Sert. C_3 No. 3, 93 (1964).

9) *Northrop, J. H., M. Kunitz*, J. Gen. Physiol. **7**, 929 (1925).

10) *Overbeek, J. Th. G.*, cited in: Adv. in Colloid Sci. Vol. 3, p. 97 (New York, 1950).

11) *Ottewill, R. H.* et al., Trans. Faraday Soc. **56**, 854 (1960).

12) *Verwey, E. J. W., D. Th. G. Overbeek*, Theory of stability of lyophobic Colloids (Amsterdam 1948).

13) *Derjaguin, B.*, J. Expt. Thur. Phys. URSS **11**, 802 (1941).

14) *Derjaguin, B., L. D. Landau*, Acta Physicochem. URSS **14**, 633 (1941).

15) *Derjaguin, B., M. Kussakov*, Acta Phys. Chem. URSS **10**, **25**, 125 (1939).

16) *Schenkel, J. N., J. A. Kitchener*, Trans. Faraday Soc. **56**, 161 (1960).

17) *Sastry, T. G., S. N. Srivastava*, Bull. Chem. Soc. Japan **42**, 291 (1969).

18) *Sharma, M. K.* et al., Ind. J. Technol. **13**, 419 (1975).

Authors' address:

M. K. Sharma
Chemical Laboratories, Agra College
Agra-282002 (India)

Progr. Colloid & Polymer Sci. 63, 65–67 (1978)

ISSN 0340-255 X

Chemistry Department, Agra College, Agra (India)

Flocculation of oil-in-water emulsions by detergents

P. Lata and *P. Bahadur*

With 2 figures and 1 table

(Received March 29, 1976)

Introduction

Detergents have long been used in the stabilization of emulsions (1, 2) and these were found probably the most efficient emulsifiers (3, 4). Despite their emulsifying action, we here found, that these materials can be used, equally well, in breaking the emulsions.

Materials and methods

The detergents used were:

Anionic
(i) Dibutyl ester of sodium sulpho succinic acid (BSS),
(ii) Dioctyl ester of sodium sulpho succinic acid (OSS),
(iii) Sodium dodecyl benzene sulphonate (SDBS),
(iv) Sodium alkyl sulphate containing 17 carbon atoms (SAS)

Cationic
(i) Tetradecyl pyridinium bromide (TPB),
(ii) Hexadecyl pyridinium bromide (HPB),
(iii) Hexadecyl pyridinium chloride (HPC),
(iv) Hexadecyl trimethyl ammonium bromide (HTAB)

All the detergents were BDH chemicals. Toluene used was AnalaR grade. Double distilled water from all pyrex glass apparatus was used throughout the measurements.

Toluene-in-water emulsions were prepared by suspending 5% toluene in an aqueous solution of 0.001 M emulsifier in 0.01 M KCl. The KCl was added to maintain the value of Debye Hückel parameter almost constant. The mixture was hand-shaken for about fifteen minutes and then passed twice through a hand operated stainless steel homogenizer which resulted the emulsion droplets of the diameter ranging 1–2 microns.

The emulsion was kept for some time and the creamed layer was skimmed off. This was then diluted ten times maintaining the same ionic strength by the detergent solutions of different concentration. The zeta potentials were calculated micro-electrophoretically using a Northrop-Kunitz type cell which was placed in lateral position. The microscope was fixed at stationary level. The zeta potential values were reproducible to 4%.

Detergents themselves were used as emulsifiers in preparing stable emulsions. Sodium alkyl sulphate formed the emulsion with negatively charged droplets whereas HTAB formed positively charged emulsions. Thus the flocculation of these emulsions has been critically examined by the detergents.

Results and discussion

Both the SAS and HTAB stabilized emulsions were very stable because of appreciable charge on the emulsion globules. Microscopic observations revealed the decrease in particle number on aging and also that the flocculation is the major factor controlling the stability. The observed flocculation was thus accounted for as reversible by nature occurring in the secondary minimum region. The occurrence of primary minimum flocculation seemed here improbable because of very high energy humps of even longer than about thousands kT. Similar flocculation was discussed, on the quantitative basis in previous communications (3, 4, 5).

For SAS stabilized emulsions where the droplets were negatively charged the flocculating effect of cationic detergents was observed. Normally, the decrease in zeta potential was seen. Initially this decrease was very low which was followed by a sudden fall in the zeta potential upto the isoelectric point of the system. On further addition of cationic detergents an increase in zeta potential, though in negative direction, was noticed which was followed by constancy. The zeta potential-log C curves are shown in figure 1. The flocculating concentrations of detergents are reported in table 1. It was observed from the data that all the detergents were capable of

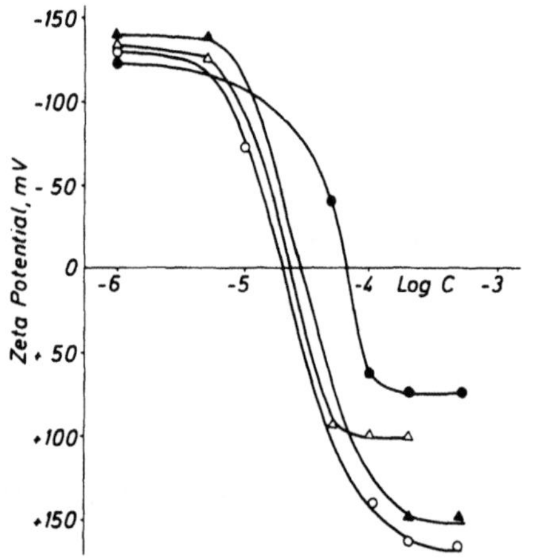

Fig. 1. Zeta potential-log C curves for SAS stabilized emulsions. ● TPB, △ HTAB, ▲ HPB, ○ HPC

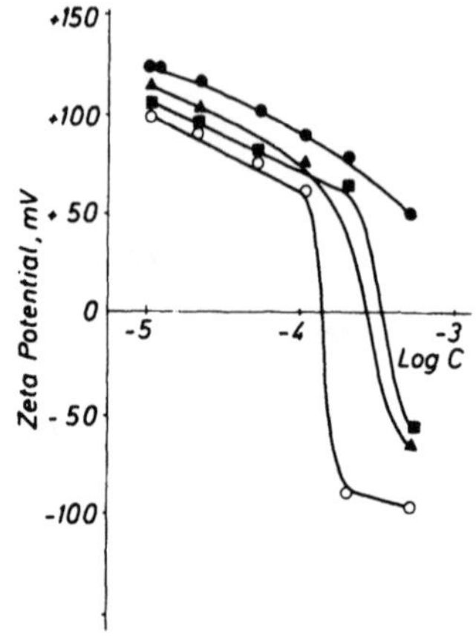

Fig. 2. Zeta potential-log C curves for HTAB stabilized emulsions. ● BSS, ■ SDBS, ▲ SAS, ○ OSS

Table 1. Flocculating Concentrations of Detergents

Emulsion Stabilizer	Flocculating Detergent	Flocculating Concentration (Moles/litre)
SAS	TPB	6.31×10^{-5}
	HPB	2.37×10^{-5}
	HTAB	2.24×10^{-5}
	HPC	2.00×10^{-5}
HTAB	SAS	3.55×10^{-5}
	SDBS	3.17×10^{-5}
	OSS	1.89×10^{-5}

bringing the system at isoelectric point at concentrations even less than 10^{-5} M. These values were comparatively lower than those observed with polyvalent inorganic salts. The chain length of the detergents had pronounced effect in flocculation of the system. The sufficiently high values of TPB than HPB show the chain length effect. However, the group was not found to show any effect.

For HTAB system, the flocculating effect of various anionic detergents was observed. All the detergents except BSS reversed the charge of the system. The values of flocculating concentration shows the relative efficacy of these materials. For detergents with greater chain length, the flocculation effect was more pronounced. The uncommon behaviour of BSS was because of its small carbon chain length. The zeta potential-log C curves for these detergents are shown in figure 2.

Microscopic studies revealed the continuous decrease in particle number with the concentration of the detergents. After the charge reversal the system again acquired stability, because of regaining the charge by the adsorption of excess of effective ions on the globules. However, in such cases the particles were bigger in size since both the flocculation and coalescence had occurred at the time of charge reversal. For the detergent concentrations at which the isoelectric point of system was close, the emulsion was completely broken and no individual droplets were seen.

Acknowledgment

One of the authors (P. B.) is thankful to CSR for the award of a Post Doctoral Research Fellowship.

Summary

Ionic detergents were used in flocculating various toluene-in-water emulsions. These were found even more capable than polyvalent ions in their flocculation effect and only a very small amount was needed to cause the complete flocculation. The flocculation of negatively charged emulsion was governed by cationic detergents whereas for positively charged emulsions, the anionic detergents were used as flocculating agents.

Zusammenfassung

Zur Flockung von Toluol-in-Wasser-Emulsionen wurden ionische Detergentien verwendet. Diese erwiesen sich in ihrer flockenden Wirkung als wirkungsvoller als mehrwertige Ionen. Es wurden nur sehr geringe Mengen benötigt, um vollständige Flockung zu erreichen. Für negativ geladene Emulsionen waren kationische Detergentien, für positiv geladene Emulsionen anionische Detergentien die flockenden Agentien.

References

1) *Becher, P.*, Emulsions (New York 1965).
2) *Sherman, P.*, Emulsions Science (New York 1968).
3) *Bahadur, P., S. N. Srivastava*, Indian J. Technol. **12**, 343 (1974).
4) *Bahadur, P., S. N. Srivastava*, Indian J. Technol. **13**, 181 (1975).
5) *Singh, L. R., P. Bahadur, S. N. Srivastava*, J. Colloid Pol. Sci. (Communicated).

Authors' address:

P. Lata and *P. Bahadur*
Chemistry Department,
Agra College
Agra 282002 (India)

Progr. Colloid & Polymer Sci. **63**, 68–74 (1978)

ISSN 0340-255 X

Chemistry Department, B. S. A. College, Mathura (India)

Electrolyte flocculation of oil-in-water emulsions stabilized by anionic detergents

*L. R. Singh, P. Bahadur**), and *P. Lata**)

With 9 figures and 1 table

(Received March 29, 1976)

Introduction

The adsorbability of ions by colloid particles and their influence on coagulation was investigated by *Freundlich* (1). He established the relation between the coagulating powers of oppositely charged ions and their adsorbabilities. According to him the phenomena of coagulation would be governed mainly by the preferential adsorption of the counter ions and neutralization of charge. He also studied the effect of electrolyte concentration on the electrokinetic potential of colloidal dispersions. It was observed, in general, that there was first of all an increase in the value of zeta potential which followed its fall to zero value on further addition of electrolyte. *Hardy* (2) suggested that on addition of electrolytes the charge on particles is reduced by the adsorption of oppositely charged ions and the stability of sol becomes least at isoelectric point.

The effect of electrolytes to soap stabilized emulsions was investigated by *Clowes* (3) and *Bhatnagar* (4). A clear indication of possibility for influencing the flocculation and coalescence separately was discussed by *Elkar* et al. (5). *Kremnev* (6) studied the influence of electrolytes on the thickness of the interfacial film of a soap stabilized benzene-in-water system. *Van den Tempel* (7) showed that the coalescence occurred just after breaking of the film separating the oil globules in contact. Similar studies were also made by *Gillespie* and *Rideal* (8). *Wellman* and *Tartar* (9) showed that polyvalent metal soaps of small chain fatty acids always favour oil-water type emulsions and did not stabilize water-oil emulsions.

Van den Tempel (10) measured the rate of coagulation of anionic soaps stabilized oil-in-water emulsions and observed an increase in emulsion stability effects with the increase in contents of emulsifiers. *Sastry* (11) investigated, hemocytometrically, the effect of electrolytes on oil-in-water emulsions and observed that flocculation and coalescence increased with the concentration of the added electrolyte. Recently *Jain* and *Srivastava* (12) also noticed the decrease in charge on colloidal particles on the basis of thickness of the double layer in presence of varying amounts of electrolytes.

The present paper deals with the effect of some electrolytes viz. thorium nitrate, chromium nitrate, uranyl nitrate, lead nitrate and sodium nitrate on the flocculation of benzene-in-water emulsions stabilized by anionic surfactants. Attempts have also been made to apply the DLVO (13–15) theory to the flocculation results in order to discuss the stability of such emulsions in quantitative manner. The calculations of interaction energies and flocculating concentrations are made by using three important variables namely particle size 'a', zeta potential and Debye-Hückel parameter $\varkappa$.

Experimental

The surfactants used were all anionic in nature. These were dibutyl ester of sodium sulphosuccinic acid (BSS), dioctyl ester of sodium sulphosuccinic acid (OSS) and sodium salt of secondary alkyl sulphate containing seventeen carbon atoms (SAS). Thorium nitrate, chromium nitrate, lead nitrate and sodium nitrate used were all BDH AnalaR grade. Uranyl

*) Chemistry Department, Agra College, Agra (India).

nitrate was M & B quality. Benzene and potassium chloride used were also of BDH AnalaR grade.

Ordinary distilled water was redistilled from all pyrex apparatus and this doubly distilled water was used throughout the experiments. All glass-ware were thoroughly cleaned before use. The emulsions of different surfactants were prepared by dispersing 5% benzene in aqueous 0.01 M KCl solution containing 1 mM of surfactant. The electrophoretic mobility and zeta potential of emulsion droplets in presence of different amounts of flocculating agents were carried out by microelectrophoresis apparatus. The zeta potentials were calculated by Hückel equation. The size of emulsion droplets was determined by microphotographic technique.

Results and discussion

A. The particle size and concentration

For the size of emulsion droplets a drop of freshly prepared emulsion was taken on a hemocytometer (German double Neubauer model) under an Olympus microscope attached with an Olympus camera; with the help of which microphotographs of emulsions were taken at a magnification 15 × 40. These microphotographs were further enlarged two times which finally resulted in a magnification of 1200 times. The particle size of the emulsion droplets were determined for each microphotograph with the help of a cathetometer by counting about 200 particles and plotting size frequency histograms. The average particle radii found in such a way were 1.2, 1.1 and 1.0 microns for BSS, SAS and OSS stabilized emulsions respectively. A representative histogram is shown in figure 1.

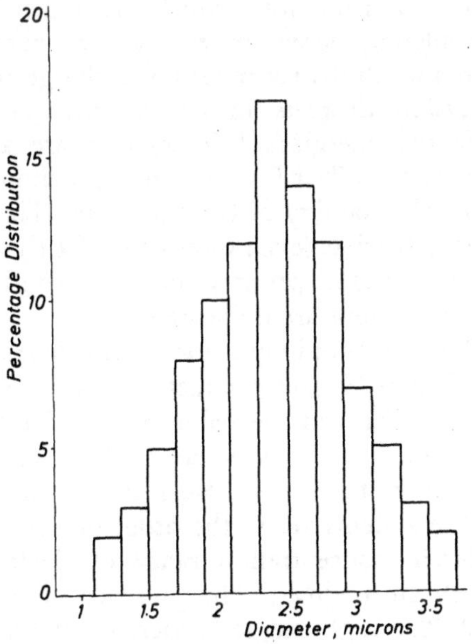

Fig. 1. The size frequency curve for BSS stabilized emulsion

For the determination of the initial particle concentration for the systems, a drop of freshly diluted emulsion was placed on a hemocytometer and was examined under an Olympus microscope and the number of droplets in 50 squares of a calibrated graticule was counted with the help of a tally counter. This was finally converted as the particles per ml. From these observations the initial number of particles present in 1 ml of emulsion was determined and found to be 6.8×10^8, 8.2×10^8 and 9.7×10^8 particles/ml for BSS, SAS and OSS emulsions respectively.

B. The zeta potential

The variations in the values of zeta potential of the emulsion droplets by the addition of different amount of inorganic salts with different metal cations were studied. The initial zeta potential values of the fresh emulsions were 105.0 mV, 145.0 mV and 163.2 mV of BSS, SAS and OSS systems respectively. The ζ-log C curves in presence of nitrates of thorium, chromium, uranyl, lead and sodium are shown in figures 2–4 for all three stabilized emulsions. In all cases, addition of electrolyte reduces the charge on the emulsion droplets and it was found that only tetravalent thorium ions were capable of bringing the systems at isoelectric point under the concentration range studied. The flocculating concentrations of $Th(NO_3)_4$ were 1.844×10^{-5} M, 7.079×10^{-5} M and 1.413×10^{-4} M for BSS, SAS and OSS stabilized emulsions respectively which indicate the stabilizing powers of the surfactants in the order OSS > SAS > BSS. The similar order of efficacy has been observed from the adsorption data (16). After the isoelectric point, the charge on the emulsion droplets again increases but in the reverse direction with the further addition of $Th(NO_3)_4$ and reaches upto about 50 mV and then acquires constancy. Moreover, the order of flocculating powers of metal cations was noticed in accordance with Hardy-Schulze rule which seems $Th^{+4} > Cr^{+3} > UO_2^{+2} > Pb^{+2} > N^{+1}$ for all the systems.

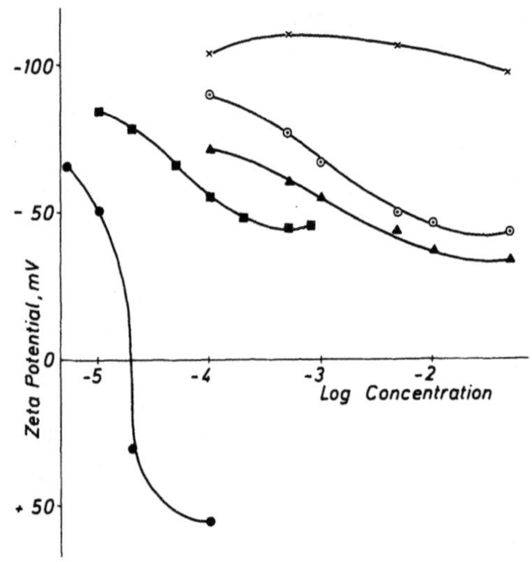

Fig. 2. Zeta potential – log C curves for BSS system: × $NaNO_3$, ○ $Pb(NO_3)_2$, ▲ $UO_2(NO_3)_2$, ■ $Cr(NO_3)_3$, ● $Th(NO_3)_4$

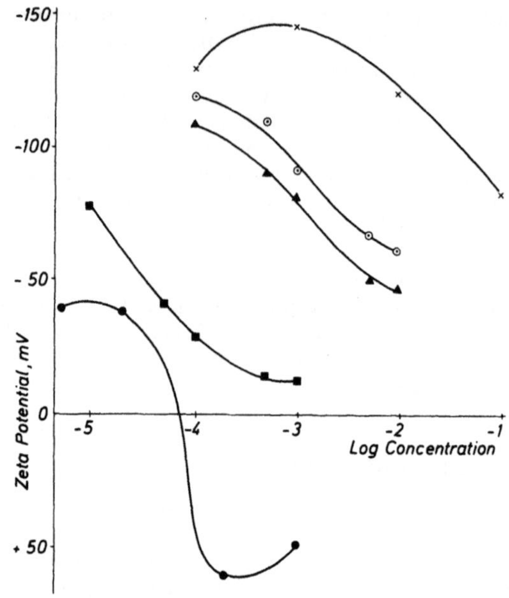

Fig. 4. Zeta potential – log C curves for SAS system: × $NaNO_3$, ⊙ $Pb(NO_3)_2$, ▲ $UO_2(NO_3)_2$, ■ $Cr(NO_3)_3$, ● $Th(NO_3)_4$

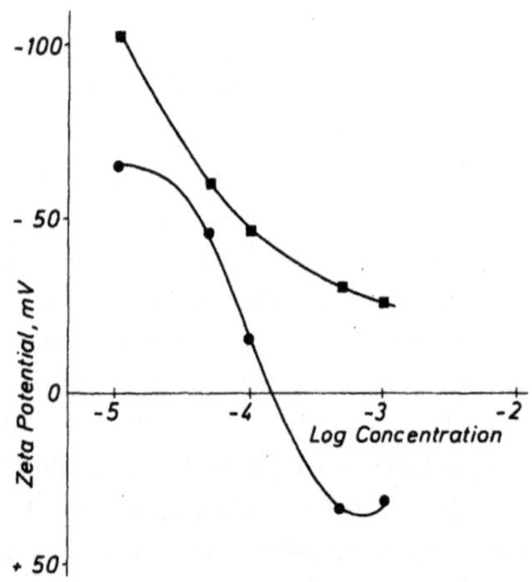

Fig. 3. Zeta potential – log C curves for OSS systems: ■ $Cr(NO_3)_3$, ● $Th(NO_3)_4$

C. The interaction energy curves and emulsion stability

The interaction energies of the double layer around the particle for different electrolyte concentrations were calculated by taking the algebraic sum of potential energy of repulsion (17) and attraction (18). The values of van der Waals constant was theoretically computed for pure benzene dispersed in water which came out 2.05×10^{-13} erg. The energy diagrams for BSS systems only are shown in figures 5-9. The values of zeta potential, Debye-Hückel parameter, height of maxima, depth of secondary minima etc. are shown in table 1 for all the three emulsions. From the interaction energy curves showing the energy maxima and depth of secondary minima, a few common conclusions can be drawn. It has been observed that energy maxima decrease generally with the addition of salt. Only a sharp change in energy barriers has been observed for tetravalent thorium ions which are efficient for flocculating the systems even at low concentration and ultimately reverse the charge on the emulsion droplets in all systems. The net potential energies of interaction are always greater than 25 kT which are high enough to show the stability of the emulsions. The high energy barriers (even thousands of kT) would obviously avoid primary flocculation and show that emulsions are quite stable.

The extent of interaction energy depends on the Debye-Hückel parameter, zeta potentials, particle size and the value of van der Waals constant, which vary with the electrolyte content of the system. Here also $\varkappa$ values for all the systems are in the order of 10^6–10^7 at different concentration range of electrolytes which are owing to the presence of 0.01 M KCl. The high energy barriers imply the high degree of stability of emulsions but in practice the flocculation does occur which can be

Table 1. Values of Debye-Hückel parameters, zeta potential, energy maxima and depth of secondary minima etc. for surfactant stabilized emulsions

Emulsifier	Electrolyte Conc. M	$\varkappa \times 10^{-6}$	Zeta potential mV	Energy Maxima kT	Inter-particle distance at $V=0$ Å	Depth of secondary minima kT
SAS	$NaNO_3$					
	1×10^{-4}	3.261	−128.30	9900	–	–
	1×10^{-3}	3.403	−145.00	12750	–	–
	1×10^{-2}	4.590	−119.00	7907	114	18.5
	1×10^{-1}	10.760	− 81.06	2410	32	30.4
	$Pb(NO_3)_2$					
	1×10^{-4}	3.294	−117.90	8300	–	–
	5×10^{-4}	3.479	−108.30	6750	196	4.8
	1×10^{-3}	3.700	− 89.99	4355	157	11.0
	5×10^{-3}	5.133	− 65.75	1825	103	19.5
	1×10^{-2}	6.489	− 60.41	1330	68	20.2
	$UO_2(NO_3)_2$					
	1×10^{-4}	3.294	−106.90	6560	–	–
	5×10^{-4}	3.479	− 88.53	4240	182	6.0
	1×10^{-3}	3.700	− 81.21	3410	154	11.5
	5×10^{-3}	5.133	− 49.50	820	93	19.5
	1×10^{-2}	6.489	− 47.33	830	52	20.0
	$Cr(NO_3)_3$					
	1×10^{-5}	3.255	− 77.02	3090	195	5.0
	5×10^{-5}	3.294	− 41.30	650	130	9.8
	1×10^{-4}	3.342	− 28.02	180	105	14.0
	5×10^{-4}	3.700	− 14.13	–	–	–
	1×10^{-3}	4.104	− 11.63	–	–	–
	$Th(NO_3)_4$					
	5×10^{-6}	3.251	− 39.22	548	130	10.0
	2×10^{-5}	3.277	− 39.03	532	125	9.0
	Charge reversal					
	2×10^{-4}	3.550	60.95	1700	147	12.0
	1×10^{-3}	4.590	48.93	795	103	17.2
OSS	$Cr(NO_3)_3$					
	1×10^{-5}	3.255	−102.00	5940	–	–
	5×10^{-5}	3.295	− 58.82	1580	172	6.5
	1×10^{-4}	3.342	− 46.25	880	128	10.8
	5×10^{-4}	3.700	− 30.62	260	110	13.4
	1×10^{-3}	4.104	− 26.36	120	82	18.0
	$Th(NO_3)_4$					
	1×10^{-5}	3.261	− 63.87	2062	183	6.5
	5×10^{-5}	3.326	− 45.74	825	145	8.5
	1×10^{-4}	3.403	− 15.49	–	–	–
	Charge reversal					
	5×10^{-4}	3.975	− 33.64	290	103	17.2
	1×10^{-3}	4.590	29.86	145	73	19.0
BSS	$NaNO_3$					
	1×10^{-4}	3.261	−103.40	6155	–	–
	5×10^{-4}	3.326	−110.60	7100	–	–
	5×10^{-3}	3.975	−105.80	6300	151	12.4
	5×10^{-2}	7.950	− 96.90	3375	75	21.5

Table 1. (continued)

Emulsifier	Electrolyte Conc. M	$\varkappa \times 10^{-6}$	Zeta potential mV	Energy Maxima kT	Inter-particle distance at $V=0$ Å	Depth of secondary minima kT
	$Pb(NO_3)_2$					
	1×10^{-4}	3.294	– 87.80	–	–	–
	5×10^{-4}	3.479	– 77.32	3060	–	–
	1×10^{-3}	3.700	– 65.89	2025	147	16.5
	5×10^{-3}	5.133	– 48.73	805	94	19.8
	1×10^{-2}	6.489	– 45.53	540	57	20.0
	5×10^{-2}	12.980	– 42.71	90	18	49.8
	$UO_2(NO_3)_2$					
	1×10^{-4}	3.294	– 70.82	2500	185	5.5
	5×10^{-4}	3.479	– 58.84	1570	142	8.5
	1×10^{-3}	3.700	– 54.42	1260	133	13.0
	5×10^{-3}	5.133	– 43.02	540	87	19.5
	1×10^{-2}	6.489	– 36.24	225	50	22.5
	5×10^{-2}	12.980	– 33.44	–	–	–
	$Cr(NO_3)_3$					
	1×10^{-5}	3.255	– 83.62	3750	–	–
	2×10^{-5}	3.266	– 76.90	3055	194	7.0
	5×10^{-5}	3.294	– 66.49	2175	180	8.0
	1×10^{-4}	3.342	– 54.14	1250	137	14.0
	2×10^{-4}	3.434	– 47.22	828	120	16.5
	5×10^{-4}	3.700	– 43.76	460	57	21.0
	8×10^{-4}	3.947	– 45.22	325	27	28.0
	$Th(NO_3)_4$					
	5×10^{-6}	3.251	– 65.09	2170	175	4.9
	1×10^{-5}	3.261	– 50.01	1060	151	6.5
	Charge reversal					
	2×10^{-5}	3.277	30.00	235	120	12.5
	1×10^{-4}	3.403	53.88	1260	155	6.5

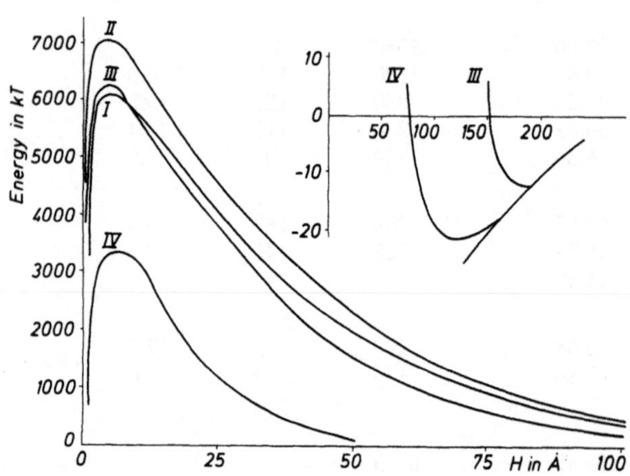

Fig. 5. The energy diagram for BSS system flocculated by $NaNO_3$. I. 1×10^{-4}; II. 5×10^{-4}; III. 5×10^{-3}; IV. 5×10^{-2} M

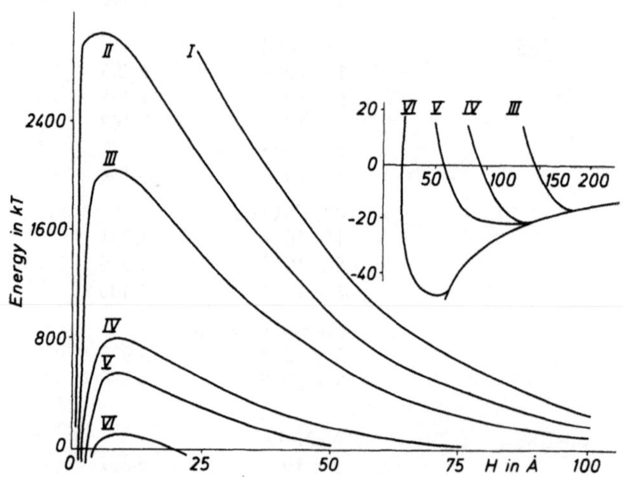

Fig. 6. The energy diagram for BSS system flocculated by $Pb(NO_3)_2$. I. 5×10^{-4}; II. 1×10^{-3}; III. 5×10^{-3}; IV. 1×10^{-2}; V. 5×10^{-2} M

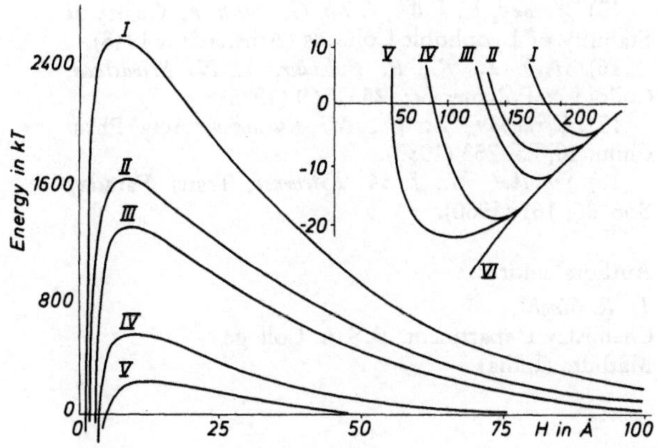

Fig. 7. The energy diagram for BSS system flocculated by $UO_2(NO_3)_2$. I. 1×10^{-4}; II. 5×10^{-4}; III. 1×10^{-3}; IV. 5×10^{-3}; V. 1×10^{-2}; VI. 5×10^{-2} M

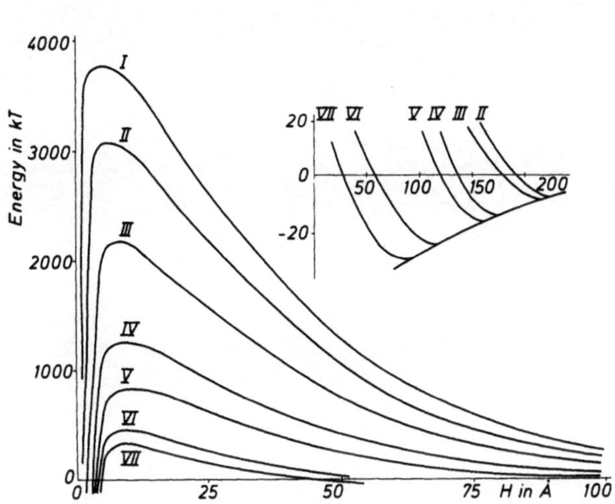

Fig. 8. The energy diagram for BSS system flocculated by $Cr(NO_3)_3$. I. 1×10^{-5}; II. 2×10^{-5}; III. 5×10^{-5}; IV. 1×10^{-4}; V. 2×10^{-4}; VI. 5×10^{-4}; VII. 8×10^{-4} M

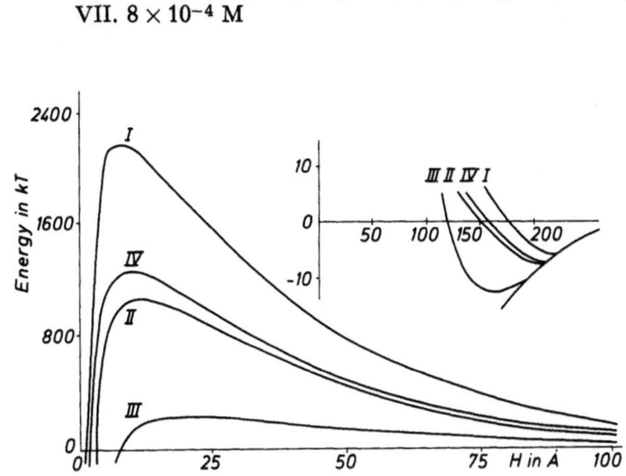

Fig. 9. The energy diagram for BSS system flocculated by $Th(NO_3)_4$. I. 5×10^{-6}; II. 1×10^{-5}; III. 2×10^{-5}; IV. 1×10^{-4} M

explained only if secondary minimum aggregation is assumed. There is always fall in interaction energy with the interparticle distance and at larger distances attraction may predominate. It is also confirmed from the interaction energy curves of all the systems that at larger distances the attraction predominate and the secondary minima are present. The enlarged secondary minima portions are shown along side of the energy figures. It is clear from the figures that depth of secondary minima are reasonably enough to trap the droplets and thereby cause flocculation.

Summary

The flocculation of negatively charged benzene-in-water emulsions was governed by various metal cations viz., Na^{1+}, Pb^{2+}, UO_2^{2+}, Cr^{3+} and Th^{4+}. The anionic detergents BSS, OSS and SAS were the emulsifying agents. The flocculation studies were made by determining the zeta potentials microelectrophoretically. All the electrolytes except $NaNO_3$ reduced the zeta potential appreciably. However, the charge reversal was observed with $Th(NO_3)_4$ only. The flocculating powers of the salts were of the order $Th^{+4} > Cr^{+3} > UO_2^{+2} > Pb^{+2} > Na^{+1}$. The emulsifying powers of the detergents were of the order OSS > SAS > BSS. The results were interpreted in the light of DLVO theory and the flocculation was largely found to occur in secondary minima which is by nature reversible.

Zusammenfassung

Es wurde das Flockungsverhalten negativ geladener Benzol-in-Wasser-Emulsionen studiert, die mit Hilfe der anionischen Detergentien BSS, OSS und SAS als Emulgatoren hergestellt waren. Als flockende Elektrolyte wurden Na^+, Pb^{2+}, UO_2^{2+}, Cr^{3+} und Th^{4+} verwendet. Die flockende Wirkung steigt in der angegebenen Reihenfolge. Ladungsumkehr wurde nur mit Th^{4+} bewirkt. Die Emulgatorwirkung sinkt in der Reihe OSS > SAS > BSS. Die Ergebnisse wurden auf der Basis der DLVO-Theorie interpretiert. Die Flockung erfolgt weitgehend im sekundären Minimum, welches seiner Natur nach reversibel ist.

References

1) *Freundlich, H.*, Z. Physik. Chemie **57**, 462 (1907); Trans. Faraday Soc. **19**, 66 (1913).

2) *Hardy, J.*, Physiol. **24**, 288 (1899); Nature **109**, 226 (1921).

3) *Clowes, G. H. A.*, J. Phys. Chem. **20**, 407 (1916).

4) *Bhatnagar, S. S.*, J. Chem. Soc. (London) **117**, 542 (1920).

5) *Elkar* et al., Proc. Roy. Soc. (London) **58**, 243 (1939); **59**, 185 (1940).

6) *Kremnev, L. Ya., H. I. Kuibina*, Kolloid-Z. **9**, 392 (1947).

7) *van den Tempel, M.*, J. Colloid Sci. **13**, 125 (1958).

8) *Gillespie, T., E. K. Rideal*, J. Colloid Sci. **10**, 281 (1955).
9) *Wellman, V. E., H. V. Tartar*, J. Phys. Chem. **34**, 370 (1930).
10) *van den Tempel, M.*, Rac. Trav. Chim. **72**, 442 (1953).
11) *Sastry, T. G.*, Dissertation (Agra 1969).
12) *Jain, S. P., S. N. Srivastava*, Ind. J. Chem. **8**, 731 (1970).
13) *Derjaguin, B. V.*, Trans. Faraday Soc. **26**, 203 (1940).
14) *Derjaguin, B. V., L. D. Landau,* Acta. Phys. Chim. **14**, 633 (1941).
15) *Verwey, E. J. W., J. Th. G. Overbeek*, Theory of Stability of Lyophobic Colloids (Amsterdam 1948).
16) *Singh, L. R., P. Bahadur, S. N. Srivastava*, Colloid & Polymer Sci. **253**, 769 (1975).
17) *Derjaguin, B. V., M. Kussakov*, Acta Phys. Chim. **10**, 25, 253 (1939).
18) *Schenkel, M., J. A. Kitchener*, Trans. Faraday Soc. **56**, 161 (1960).

Authors' address:
L. R. Singh
Chemistry Department, B. S. A. College
Mathura (India)

Progr. Colloid & Polymer Sci. **63**, 75–77 (1978)

ISSN 0340-255 X

Instruments Research and Development Establishment, Dehradun (India)

Ultracentrifugal stability of oil-in-oil emulsions

M. K. Sharma

With 2 figures

(Received Juli 12, 1976)

Introduction

Demulsification of emulsions occurs as a two-stage process (1). In the first stage, flocculation, the droplets of the disperse phase form aggregates in which the drops have not lost their identity (such aggregation is often reversible). In the second stage, termed coalescence, each aggregate combines to form a single drop. This is an irreversible process, leading to a decrease in the number of dispersed droplets and finally to the ultimate separation of dispersed droplets as a bulk phase, i.e., so-called complete breaking of emulsion.

In continuation of earlier work (2–4), the present investigation reports the effects of initial surfactant concentration and particle size on the ultracentrifugal stability of oil-in-oil emulsions prepared with benzene and ethylene glycol as two nonaqueous phases and sodium dioctylsulphosuccinate as emulsifier.

Ultracentrifugation (5, 6) is found to offer a means for accomplishing the flocculation step very rapidly and thus producing a highly concentrated emulsion before appreciable coalescence can occur. The coalescence process then could be studied independently. In this communication emulsion stability is used exclusively to refer to the rate of coalescence.

Experimental

Preparation of Emulsions: – A series of oil-in-oil emulsions of benzene (BDH) in ethylene glycol (BDH) was prepared with various concentrations of emulsifier. A 1:1 phase ratio by volume was used and the concentration of sodium dioctylsulphosuccinate (BDH) was varied from 1 to 5% (w/v % of ethylene glycol). The heterogeneous mixtures of benzene and ethylene glycol containing anionic surfactant were emulsified with the help of Braun emulsator for the final making of the emulsions. Care was taken to reproduce conditions exactly in preparing the O/O emulsions.

Ultracentrifugal Study: – A Beckman Spinco Model E analytical ultracentrifuge was used in this work with a 12 mm. light path and a 4 degree sector angle standard cell having a capacity of about 0.8 ml. Immediately after preparation of a nonaqueous emulsion, 0.7 ml. of the emulsion was injected into the ultracentrifuge cell through a hypodermic needle. The cell was then placed in an analytical rotor and subsequently centrifuged in the ultracentrifuge at 25,980 r.p.m. The time required for acceleration to this constant speed was 2 min. and 45 seconds ± 15 seconds. Zero time in the experiments was taken as the time at which constant speed was reached. All runs were made at $30° \pm 0.1$ °C.

During centrifugation three layers develop in the cell: transparent layer of benzene, the remaining opaque, concentrated stable emulsion and the transparent layer of ethylene glycol. Photographic records were made of the cell at regular time intervals during the experiment. The layers give sharp boundaries on a photographic plate which change position with time as more benzene is separated from the emulsion. Determination of the positions of the boundaries on a series of such plates taken at successive time intervals permits calculation of the rate of separation of benzene from the emulsion. Measurements were made with a travelling microscope determining the distance from the top of the cell to the benzene-emulsion and emulsion-ethylene glycol interfaces, and to the bottom of the cell.

The volume, V, corresponding to a length, l, in the cell is given by

$$V = \frac{4}{360} \wedge t[(d+l)^2 - d^2]$$

$$= \frac{\wedge tl}{90}(2d+l)$$

where d is the distance from the centre of rotation to the top of the cell, and t is the thickness of the cell. A graph was plotted of the volume as a function of the measured value of l from the top of the cell using an optical enlargement of 2.03. The plot was then used to determine the volume from the differences between the measured distances to the different interfaces.

The per cent of the initially emulsified benzene which has separated at any given time is determined by dividing the volume of the benzene layer by the sum of the volumes of the benzene layer and the creamed emulsion. The per cent of the initially emulsified benzene which has separated then is plotted against the time of ultracentrifugation. The rate of separation at the known speed is determined from the slope of the linear portion of the curve. The slope of the line, i.e., the rate of separation of benzene in per cent per minute, is termed the ultracentrifugal stability, and its extrapolated intercept on the axis at zero time is the per cent benzene separated at zero time.

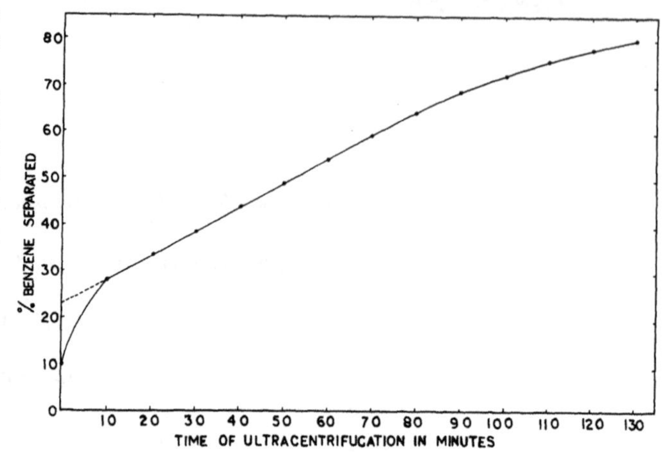

Fig. 1. Rate of separation of benzene from oil-in-oil emulsions centrifuged at 25,980 r.p.m.

Results and discussion

The boundary between the creamed emulsion and the ethylene glycol phase is formed some time within the first three minutes of centrifugation and remained in a constant position during the entire run even when there is complete coalescence of the emulsion. The sum of the volumes of clear benzene and creamed emulsion is also remained constant throughout all runs independent of the amount of benzene separated. This implies that almost all of the ethylene glycol phases of an O/O emulsion is separated from the benzene droplets very early in the centrifugation process, leaving a very concentrated emulsion consisting of benzene droplets, adsorbed emulsifier and very small amount of ethylene glycol. At this stage the system can be described as an aggregated and compressed emulsion. The droplets must assume some shape other than spherical (may be honeycomb-like shape) in order to have a dispersed phase constituting so nearly 100 per cent by volume of the concentrated emulsion. Any further decrease in the volume of this aggregated emulsion is the result of the coalescence of the benzene droplets and consequent separation into a distinct benzene phase.

Figure 1 is the characteristic of the curves of per cent benzene separated vs. time of ultracentrifugation. In this case the nonaqueous emulsion of 50% v. benzene in 50% v. ethylene glycol is stabilized with 1% w. sodium dioctylsulphosuccinate. Initially benzene separates very rapidly but the rate – determined from the slope of the curve – decreases rapidly with time during the first 10 minutes or so, during which period with 1% initial surfactant

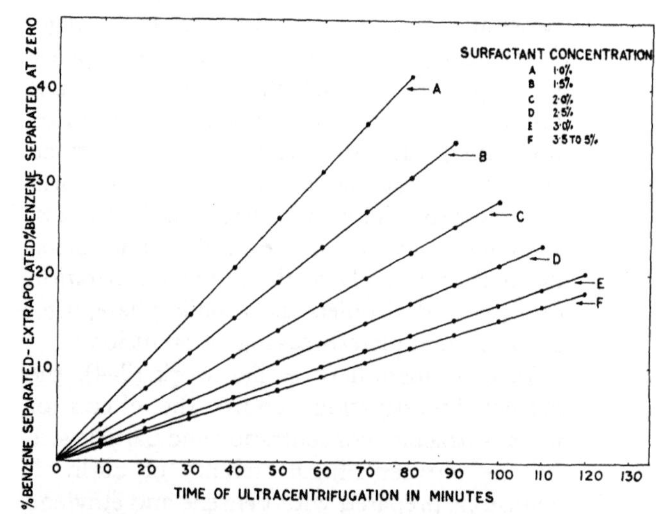

Fig. 2. Effect of initial surfactant concentration on the rate of separation of benzene from oil-in-oil emulsions centrifuged at 25,980 r.p.m.

concentration about 27% of the emulsified benzene separates as a clear layer. The rate of separation then remains constant for a considerable period until about 65% of all the benzene in the system has separated and thereafter decreases slowly with further separation of benzene.

The linear portions of the curves of emulsions prepared with different initial surfactant concentrations, are extrapolated to zero time, as shown in figure 1, and these extrapolated values of per cent benzene separated at zero time are determined.

Figure 2 shows the effect of concentration of emulsifier on the rate of separation of benzene from the emulsions in the ultracentrifuge. Increasing the concentration of surfactant at first decreases the rate of separation of benzene from the emulsion and also the amount of benzene which is separated rapidly on initial centrifugation before the steady state rate is reached. At higher concentrations the rate becomes independent of the initial concentration of emulsifier, as is strikingly shown in figure 2, all emulsions prepared with surfactant concentration between 3.5 and 5 % separating benzene at the same rate within experimental error. Figure 2 also shows the degree to which the loss of benzene by the different emulsions conforms to the straight line relationship.

From the ultracentrifugal studies, by the rate of separation of benzene from O/O emulsions, it is clear that as the initial surfactant concentration is increased, the stability of emulsions first increases and then becomes independent of concentration. In an earlier communication (4), particle size analysis of O/O emulsions showed that with an increase in emulsifier concentration upto an optimum value, the globules become smaller in size and the stability of the emulsions also increases. Thus it appears that during ultracentrifugal analysis, the rate of separation of benzene from O/O emulsions does depend on the initial size of the globules, i.e., with a decrease in the initial size of the droplets, the rate of separation of benzene from the nonaqueous emulsions decreases.

It must be emphasized that results of the ultracentrifugal method will not necessarily correlate with the shelf-life stability of the same nonaqueous emulsion since they pertain to the highly concentrated cream remaining after rapid separation of most of the ethylene glycol in the ultracentrifugal field.

It may be concluded that during ultracentrifugal experiments, the benzene droplets in oil-in-oil emulsions have been forced by centrifugation into very closely packed arrays which can be described as aggregated emulsions. The resistance of the aggregated droplets to coalescence arises from the existence of very thin ethylene glycol films between benzene droplets and from the presence of emulsifier adsorbed on the particle surfaces. The results of the experiments indicate that the rate of coalescence of emulsion droplets decreases with a decrease in the initial particle size and an increase in initial surfactant concentration until an optimum surfactant concentration is reached and thereafter is independent of concentration.

Zusammenfassung

Mit Hilfe einer Ultrazentrifuge (25.980 U/min.) wurde die Stabilität von Öl-in-Öl-Emulsionen untersucht. Die Emulsionen waren aus Benzol und Äthylenglykol als nicht mischbare Phasen und Natriumdioktylsulfosuccinat als Emulgator hergestellt. Die Koaleszenzgeschwindigkeit der Benzoltröpfchen sinkt mit abnehmender Anfangsgröße der Tröpfchen und steigender Anfangskonzentration des Tensids.

Summary

The stability of oil-in-oil emulsions prepared with benzene and ethylene glycol as two nonaqueous phases and sodium dioctylsulphosuccinate as emulsifier has been determined quantitatively employing an ultracentrifuge at 25,980 r.p.m. The effects of initial surfactant concentration and initial particle size on the ultracentrifugal stability of emulsions have been found out. The rate of coalescence of benzene droplets decreases with a decrease in initial particle size and an increase in initial surfactant concentration upto an optimum value and then becomes independent of concentration.

References

1) *van den Tempel, M.*, Rec. trav. chim. **72**, 433 (1953).

2) *Sharma, M. K.*, J. Colloid Interface Sci. **53**, 340 (1975).

3) *Sharma, M. K.*, Curr. Sci. **44**, 770 (1975).

4) *Sharma, M. K.*, Colloid & Polymer Sci. (Accepted).

5) *Vold, R. D., R. C. Groot*, J. Soc. Cosmetic Chemists **14**, 233 (1963).

6) *Garrett, E. R.*, J. Pharm. Sci. **51**, 35 (1962).

Author's address:

M. K. Sharma
Instruments Research and Development Establishment
Dehradun (U. P.)
India

Progr. Colloid & Polymer Sci. **63**, 78–86 (1978)

ISSN 0340-255 X

Department of Chemistry, University College of Science, Calcutta (India)

Electrophoresis of emulsion droplets. Application of relaxation correction and liquid drop electrophoresis equations

S. S. Jana, M. Sengupta, and *D. N. Biswas*

With 7 figures

(Received November 11, 1976)

In some earlier publications (1, 2, 3), the relaxation correction in the theory of electrophoresis of non-conducting solid spherical particles has been considered in detail. The different analytical approximations of *Overbeek* (4) and *Booth* (5), as also the computer calculation solution of the problem by *Wiersema, Loeb* and *Overbeek* (6) (W. L. O.) were considered. It was mentioned (1, 6) that mobility data for particles of different sizes in electrolyte solutions of constant concentrations are the most suitable for this purpose, because the quantity ζ is then presumably held constant. However, such data are very meagrely recorded in literature (7–10); and, generally speaking, it has not been possible to decide conclusively regarding the relative merits of the different theoretical treatments mentioned on the basis of the results of their application to the few experimental studies cited. It was the object of the present research to undertake precise measurement of electrophoretic mobilities of droplets of varying sizes of some oil-in-water emulsions (using oils of different viscosities); to study the effect thereon of the addition of counterions of different valencies at different concentrations; and to interpret the experimental results on the basis of the different theoretical treatments mentioned with a view to be able to decide between their relative merits. Since we use in our experiments liquid droplets rather than solid particles, while applying nevertheless the theoretical relaxation correction treatments pertaining to the latter, therefore the assumption is made that the liquid droplets behave as rigid particles; the results themselves show how far this is justified.

Experimental

Materials used:

(i) All glassware (pyrex) used were cleaned with hot chromic acid, and steamed before every set of experiment.

(ii) Double distilled water was used throughout (average sp. cond. = $1.3–1.6 \times 10^{-6}$ mho/cm). All experiments were done at room temperature.

(iii) Electrolytes used were all A. R. grade (KCl, $BaCl_2$, $LaCl_3$). KCl and $BaCl_2$ were taken as primary standards, whereas $LaCl_3$ was standardised by EDTA titration (Eriochrome Black-T) (11).

(iv) (a) Nujol used was spec-pure quality. (b) Iodobenzene was prepared from freshly distilled aniline by diazotisation, and the product was purified by vacuum distillation, only a small quantity of the middle fraction (b.p. 185 °C) being collected for our experiments. (c) Dimethylaniline (G. R., B. D. H.) was further purified by distilling twice with red phosphorus, and collecting the middle fraction distilling at 190–192 °C. (d) Medicinal castor-oil was purified by column chromatography (twice) over Brockmann alumina. About 10 ml of castor-oil dissolved in about 20 ml A. R. diethyl-ether was passed through an alumina column of diameter 1.2 cm and height 18 cm, and the oil was eluted by passing about 250 ml of the same solvent in small portions at a time through the column. The solvent was evaporated off slowly under suction from a water pump at room temperature. When almost all the solvent had escaped the very small amount of the still entrained ether was finally removed by passing through the oil, under constant shaking, a stream of air bubbles from a fine jet. The pure, almost odourless oil was then left overnight under vacuum.

Preparation of the emulsions:

The emulsions of the above different oils were prepared by suspending about 4–6 drops of the oil in about 500 ml double distilled water, and vigorously stirring with a high speed mechanical stirrer. The emulsions thus prepared were allowed to stand overnight before being used for mobility measurements.

For measurements in presence of different electrolytes (KCl, $BaCl_2$ and $LaCl_3$) appropriate volumes of the stock emulsions and the standard electrolyte solutions were mixed, adding double distilled water where necessary.

Micro-electrophoresis assembly and experimental procedure:

The micro-electrophoresis cell was a wide flat cell of cross section 0.252 sq. cm mounted in vertical orientation, so that convective and settling disturbances are both rendered perpendicular to the direction of the velocity gradient of the liquid in the cell (12). The overall assembly consisted of two Ag–AgCl electrodes (13) (2.0×1.5 cm each) which were connected to a source of steady high D.C. voltage, with a sensitive micro-ammeter ("Unipivot", Cambridge) connected in series. The cell assembly was fixed firmly on the stage of a Zeiss optical microscope, the stage itself being mounted in a vertical plane, so that the viewing tube, which was thus fixed perpendicularly to the electrophoresis cell, was placed horizontally. A low power objective ($20 \times$) in conjunction with an eye-piece ($28 \times$) fitted with two graticules was used: The fine micrometer (1 cm divided into 100 divisions) was used for particle size measurements, while the coarse micrometer (network, each side of the square = 1 mm) was used for mobility measurements.

The cell was first cleaned with warm chromic acid, rinsed thoroughly with distilled water and then with double distilled water, until the specific conductivity of water remained unchanged after being introduced into the cell. The cell, filled with double distilled water, was left overnight before making mobility measurements. For filling up the cell with the oil emulsion containing electrolyte of some particular concentration, it was first rinsed with the electrolyte solution of the same concentration, then filled with the particular electrolyte-emulsion mixture and left standing as such for about one hour before making mobility measurements, so that the glass walls of the cell might come to equilibrium with the experimental emulsion.

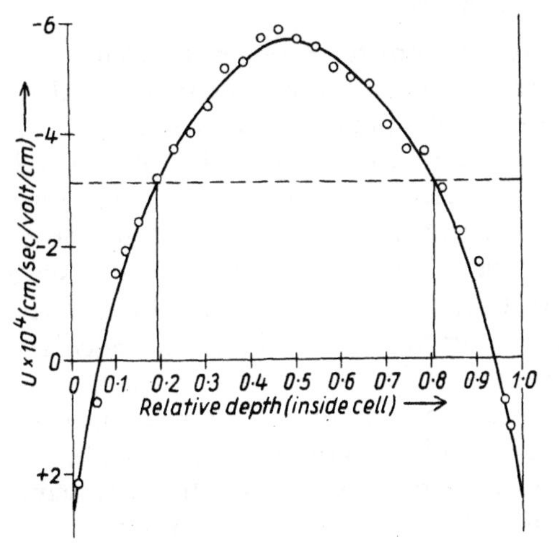

Fig. 1. Variation of mobility of Nujol droplets in distilled water with depth of position of the particle inside the microelectrophoresis cell

The profile of liquid flow inside the micro-electrophoresis cell was first traced using Nujol emulsion droplets of diameter 6.25×10^{-4} cm in distilled water (sp. cond. = 2.72×10^{-6} mho/cm), by making mobility measurements at regular intervals of depth inside the cell. The symmetric parabolic nature of the velocity profile obtained (fig. 1) showed that the cell was satisfactory in performance. Moreover, the ratio of the breadth of the cell to its inside depth being greater than 20, the so-called "stationary layers" in our cell were situated at depths of 0.21 and 0.79 times the total cell depth from either the front or the back surfaces of the cell (14). Accordingly, the microscope was fixed so that it focussed a plane situated 0,21 times cell-depth below the inner surface of the front wall of the cell, and all electrophoretic velocities were measured at this depth.

In all cases particle velocities were measured in both directions of movement; the passage of individual particles between vertical lines in the eye-piece graticule (1 div. = 52.5×10^{-4} cm) was timed, using a stop watch reading upto 0.1 sec, and the mean time noted. The diameter of the individual particles was estimated by temporarily holding it stationary in the field of the "fine" (linear) micrometer (1 small div. = 5.25×10^{-4} cm). The two micrometers were previously standardised, using a standard 'stage micrometer' properly. The specific conductivity of the suspension was measured both before and after the mobility measurements, in order to ensure that there was no significant change in conductivity on introduction of the suspension into the electrophoresis cell; only the second mentioned value was used in the calculations. The current flowing through the solution was noted.

For each electrolyte three different (low) concentrations were used. The primary object being the study of mobility at different value of $q_0 (= \varkappa a/\lambda)$, in media of constant electrolyte concentration and composition, therefore, for each electrolyte concentration mobilities were measured for particles of several different sizes. The observed size of the particles was corrected by the method described elsewhere (15), and the corrected values were used in the calculations.

Calculations and results

The particles were found to be negatively charged in all cases.

The ζ potentials were calculated (16) by using the computational method of W.L.O. This method was usefully applied only to data corresponding to comparatively low $\varkappa a$ values, otherwise no significant improvement over the ζ values calculated by the analytical approximation method of *Overbeek* would result (6). Also, it could not be applied to high ζ

systems; these are: (i) dimethylaniline emulsion in distilled water and higher valent (Ba^{+2}, La^{+3}) electrolyte solutions, except for the most concentrated $BaCl_2$ solution; (ii) iodobenzene emulsions in highest concentration Ba^{+2}, and lowest concentration La^{+3} solution; and (iii) nujol emulsion in lowest concentration $LaCl_3$ solution. However, even in those cases where ζ potential calculation by the W.L.O. method could not be carried through, the construction of the W.L.O. plots of E' vs. y_0 were nevertheless carried out in the usual manner; these curves were subsequently used for theoretical mobility calculations by the W.L.O. method, assuming characteristic constant values of ζ for these systems. The values of (i) ionic mobility parameters $m_\pm$, (ii) the $\varkappa a$ parameter q_0, (iii) the electrophoretic mobility function E', and (iv) the ζ potential function y_0, were calculated for the experimental temperatures by using (a) the usual relations (6), and (b) the corresponding values of the parameters $\lambda_\pm$, η, ϵ, etc. obtained by interpolation from the corresponding temperature – variation curves constructed by making use of literature data (17). In those cases where the W.L.O. method could not be applied, the next best method available, viz. the Overbeek analytical approximation method was used. The inverted Overbeek series for calculation of ζ potential is:

$$\zeta = \frac{U}{C_1} - \left(\frac{C_2}{C_1}\right)\left(\frac{U}{C_1}\right)^2 - \left(\frac{C_3}{C_1}\right)\left(\frac{U}{C_1}\right)^3,$$

where C_1, C_2 and C_3 are easily expressed (1) in terms of the Overbeek functions $f_1(\varkappa a)$, $f_2(\varkappa a)$, $f_3(\varkappa a)$ and $f_4(\varkappa a)$, these f's being functions of (i) the product of Debye-Hückel parameter and the radius of the particle ($\varkappa a$), (ii) the equivalent conductances of the ions present in solution ($\lambda_\pm$), (iii) the dielectric constant ϵ, viscosity coefficient η and temperature of the medium. The values of these f functions corresponding to the experimental values of $\varkappa a$ were obtained from their graphical plots (16). For sake of comparison, the ζ values were also calculated in some cases by the direct Overbeek series: $U = C_1\zeta + C_2\zeta^2 + C_3\zeta^3$. However, for this the Hunter expression (18) was not used; rather, a method of solution of the cubic equation by use of the De-Moivere's theorem was built up (16).

Among the actual details of calculations by the W.L.O. (6) method the only points which need be cited are: (i) The fact that the particles were negatively charged was duly taken into account in the calculations (2, 16); (ii) The appropriate E' vs. q_0 curves (6) were used for the 1–1, 2–1 and 3–1 electrolyte solutions; (iii) Ionic mobility correction, arising due to the deviation of the ion conductance values from 70 mho. cm², was first carried out in the case of nujol and iodobenzene emulsions, for all the different electrolytes (including KCl solution at 30 °C); the resulting corrections were however found to be insignificantly small in all cases; hence no correction was deemed necessary for dimethylaniline and castor-oil emulsions; (iv) Due to the limited range of availability of the computer calculation results [$\leqslant$ 150 mV for univalent counterions, $\leqslant$ 75 mV for 2–1 electrolytes, $\leqslant$ 50 mV for 3–1 electrolytes], the ζ potential calculations could not be carried through in the unfavourable cases of a combination of low $\varkappa a$ and high ζ values. In these cases, which have been enumerated above, the Overbeek method was used; however, this suffers from its own limitations.

The calculated ζ potential values are shown in figures 2–5 for dimethylaniline, iodobenzene, castor-oil and nujol emulsions respectively, in distilled water and in different electrolyte solutions in each case.

Theoretical mobility calculations by relaxation correction method

Apart from the ζ potential calculation discussed above, it is of interest to calculate the theoretical mobility vs. particle size (or $\varkappa a$, for constant $\varkappa$) according to the different relaxation correction methods, and to see how far these agree with the experimentally observed particle mobilities for the different systems. For this, suitable constant values of ζ have to be assumed. Now, since the calculated ζ potential values for any particular system do not remain constant (c.f. figs. 2–5), therefore the most obvious thing to do would be to take the average ζ value as the representative constant ζ for the system. However, in order to secure the best agreement of the calculated theoretical mobility curve with the experimental mobility values, a somewhat different

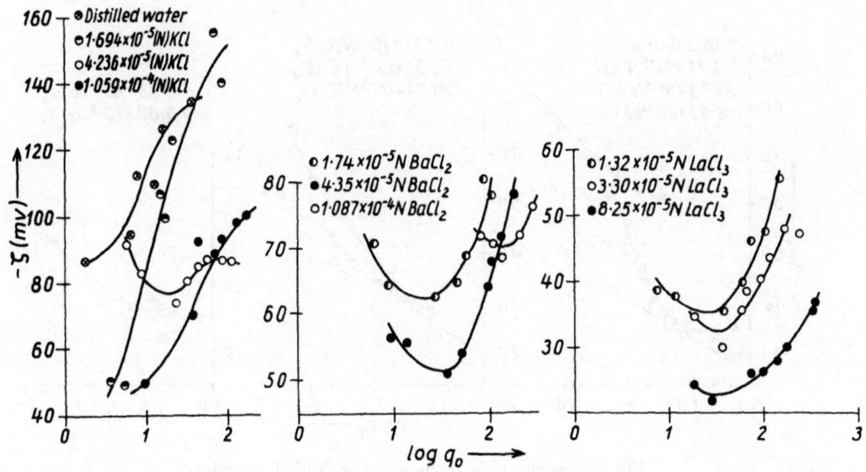

Fig. 2. ζ-potential calculation for dimethylaniline droplets in distilled water, and in various electrolytes (KCl, $BaCl_2$ and $LaCl_3$) of different constant concentrations, by the different relaxation correction methods.

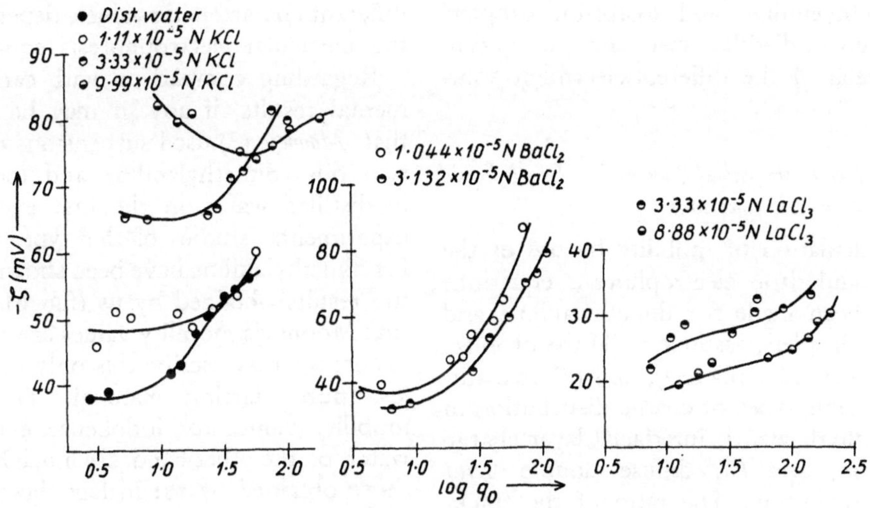

Fig. 3. Same as in figure 2 for iodobenzene droplets

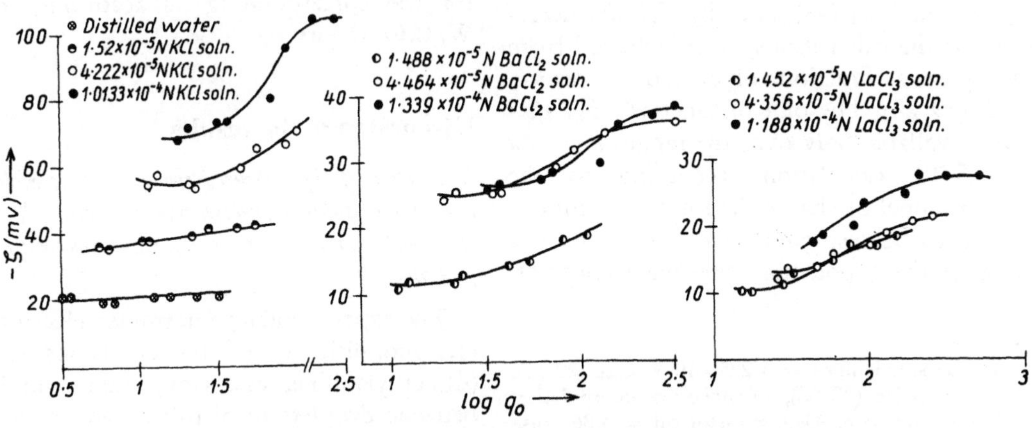

Fig. 4. Same as in figure 2 for castor-oil droplets

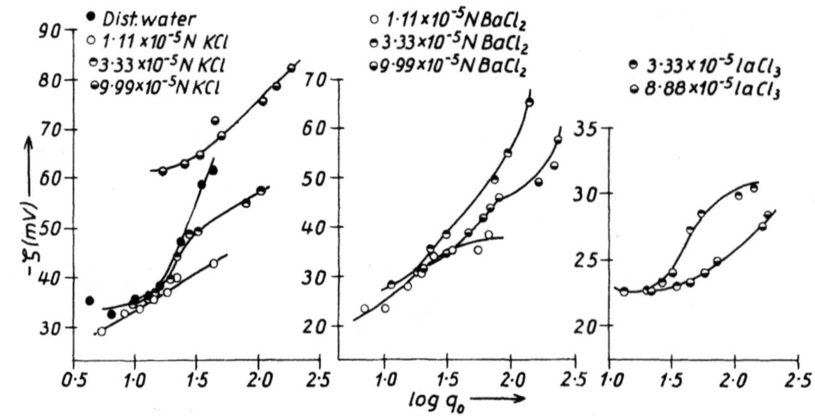

Fig. 5. Same as in figure 2 for Nujol droplets

choice of the ζ value was often found useful. The results have been shown in figures 6 and 7 for dimethylaniline and castor-oil droplets respectively in distilled water and one concentration each of the different electrolyte solutions.

Theoretical mobility calculation by the Booth liquid drop electrophoresis equations:

The calculation of mobility by use of the Booth liquid drop electrophoresis equations (19) has been made for dimethylaniline and castor-oil droplets, assuming: (1) the droplets as incompressible spheres[1]), and (2) two different possible types of charge distribution in the drop fluid, viz. (i) interfacial layer charge distribution, and (ii) diffuse double layer charge distribution. The ratio of the thicknesses of the inner and outer diffuse double layers ($\varkappa/\varkappa'$) was taken as an adjustable parameter. The values of the Booth functions (a) F_2 ($\varkappa a$), and F_1 ($\varkappa a$), and (b) F_3F_4/b', necessary for the calculations, were obtained from their plots as functions of $\varkappa a$ and $\varkappa' a$ respectively (19, 20). The plot of functions F_3F_4/b'^2 being available only over the range: $b' = \varkappa' a = 0.25$–10, calculations according to the second model of charge distribution mentioned had to be restricted over this range. The value of the ζ potential (presumed constant, under the experimental conditions) was again taken as a second adjustable parameter. It has different characteristic values, depending upon the particular electrophoresis equation used.

Regarding comparison with earlier experimental results, if any, it may be mentioned that *Mooney* (7) used suspensions of the same two oils (dimethylaniline and iodobenzene) in distilled water, in the first ever reported experimental studies of this type. His results for dimethylaniline have been shown alongside the results obtained by us (fig. 6). It is seen that Mooney's mobility values are consistently lower; we may ascribe this only to a possibly less pure starting material. The reported mobility values for iodobenzene in distilled water, on the other hand, are much higher than those obtained by us; in fact these unusually high mobility values have been commented upon by *Wiersema* (6b), who has pointed out that these exceed the theoretical limiting values for the corresponding $\varkappa a$ according to the W. L. O. theory.

Discussion of the results

Variation of (i) *measured electrophoretic mobilities and* (ii) *calculated (relaxation corrected) ζ potentials with particle size at constant electrolyte concentration:*

The experimentally determined electrophoretic mobilities of all the emulsions studied (dimethylaniline, castor-oil, nujol and iodobenzene droplets in distilled water as also in electrolyte solutions) at different constant con-

[1]) η dimethylaniline $= 1.225 \times 10^{-2}$ poise (29 °C), 1.7×10^{-2} poise (32 °C), obtained by interpolation from literature date. Also, η castor oil $= 4.86$ poise (29 °C). (Ref. 17b).

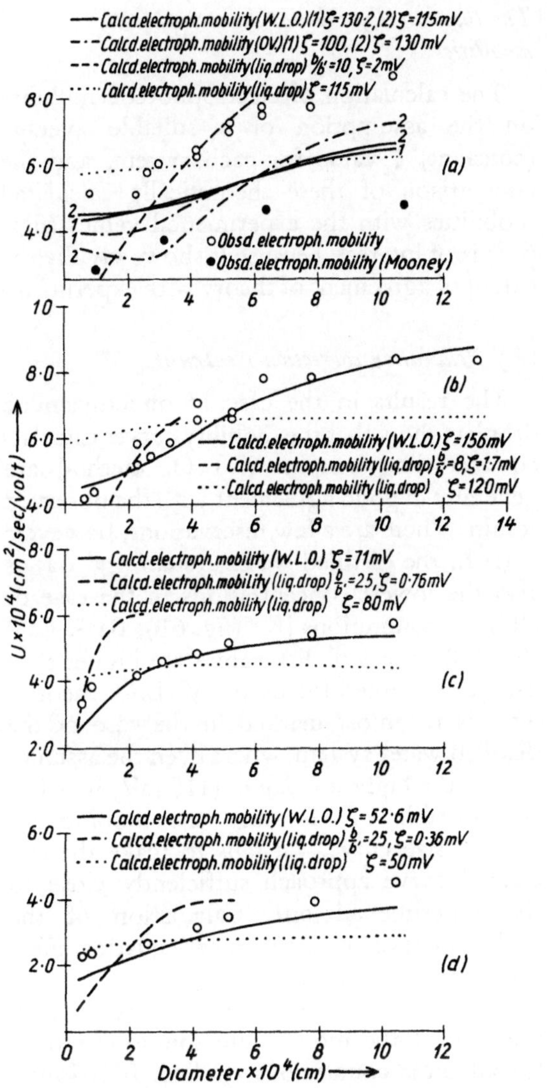

Fig. 6(a–d). Theoretically calculated electrophoretic mobilities for dimethylaniline droplets, according to the relaxation correction and the Booth liquid-drop electrophoresis equations [(a) Distilled water, (b) 1.694×10^{-5}N KCl, (c) 1.087×10^{-4}N $BaCl_2$ and (d) 1.32×10^{-5}N $LaCl_3$]

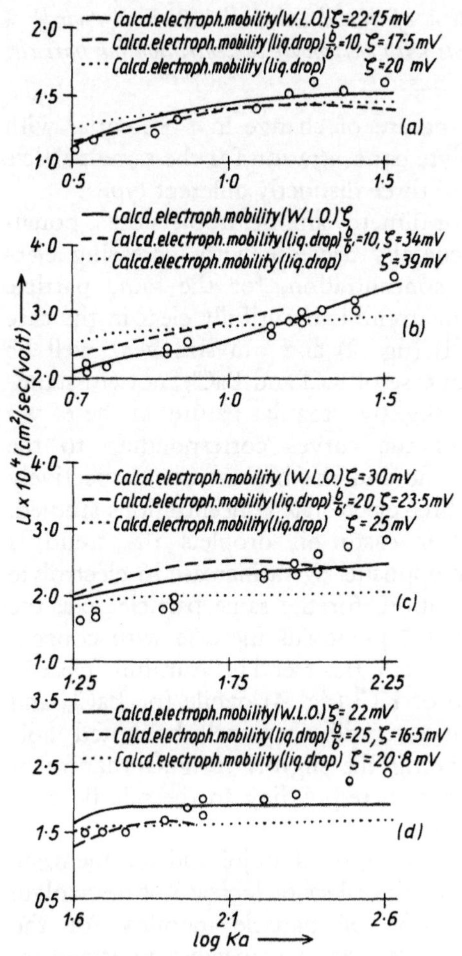

Fig. 7(a–d). Theoretically calculated electrophoretic mobilities for castor-oil droplets, according to the relaxation correction and the Booth liquid-drop electrophoresis equations [(a) Distilled water, (b) 1.52×10^{-5}N KCl, (c) 4.464×10^{-5}N $BaCl_2$ and (d) 1.188×10^{-4}N $LaCl_3$]

centrations are seen to increase steadily with increase in particle size. Earlier measurements by other workers (7, 8) have also shown an unambiguous trend of increase of electrophoretic mobility with increase in particle size in media of constant electrolyte concentration.

The calculated (relaxation corrected) ζ potentials (figs. 2–5) are found not to remain constant, but show a steady increase with particle size. This is true both of the distilled water as also of the electrolyte solution systems. The same result has also been noted in the case of Mooney's, as also *Stackelberg* et al's. electrophoretic mobility data for oil droplets (1, 2, 3).

The calculated ζ potential values show a minimum (or a tendency towards a minimum) in almost all cases in the range of low $\varkappa a$ values. This has already been noted particularly in the case of alkali solutions viz. Mooney's data for Stanolind droplets in NaOH(1) as also various organic oil droplets in KOH(2); less pronounced minima at low $\varkappa a$ values also occur in KCl solution viz., Mooney's data for organic oil droplets(3), as also *Stackelberg* et al's. data for paraffin oil droplets (3).

Variation of calculated (relaxation corrected) ζ potentials with electrolyte concentration for particles of same size:

The nature of change in ζ potential with electrolyte concentration for the same particle size is of three distinctly different types:

(i) For dimethylaniline droplets the ζ potentials generally decrease with increasing electrolyte concentration, for the same particle size. This trend is beautifully clear in the case of $LaCl_3$ (fig. 2) and can still very well be noted in case of KCl and $BaCl_2$, notwithstanding the slightly irregular nature, in these two cases, of the curves corresponding to the middle one and the highest one respectively of the three electrolyte concentrations studied.

(ii) For castor-oil droplets the trend is entirely opposite. With increase in electrolyte concentration, for the same particle size, the calculated ζ potentials increase with concentration. Again, this trend is beautifully clear in the case of KCl (fig. 4), while for $BaCl_2$ and $LaCl_3$ it can still very well be noted notwithstanding the slightly irregular nature of the curves corresponding to the middle concentration[2]).

(iii) In the case of nujol and iodobenzene emulsions, the effect of increase of electrolyte concentration on particle mobility for the same particle size is somewhat intermediate between the two different trends mentioned above; for lower valent counterions the mobility generally increases for the same particle size; however, as the counterion valency increases this trend in suppressed, and for La^{+3} as counterion the mobility eventually decreases (e.g., for the iodobenzene emulsion) for particles of same size. The change in ζ potential with the electrolyte concentration and valency of the counterions parallels the corresponding change in mobility. ζ increases with electrolyte concentration for particles of same size; this being marked for KCl and almost absent or reversed for $LaCl_3$. For iodobenzene droplets the reversal occurs already in the case of $BaCl_2$.

The theoretically calculated electrophoretic mobilities:

The calculation of electrophoretic mobility on the assumption of a suitable average (constant) ζ value for each system, and the comparison of these theoretically calculated mobilities with the experimental values (figs. 6–7) is of interest, because it shows vividly the extent of agreement of theory with experiment.

(A) *Relaxation correction treatment:*

The results in the case of dimethylamine droplets show that the mobility curves calculated according to the W.L.O. method are generally in good agreement with the observed results. There are a few reservations, however:

(i) In the distilled water system (fig. 6a) as also the lowest concentrations in the case of all three counterions [K^+ (fig. 6b), Ba^{+2}, La^{+3} (fig. 6d)], the mobility values are higher than can be accounted for by the W.L.O. theory. This is the most marked in the case of the distilled water system, where even the assumption of the highest ζ values (115 mV) which is still capable of elevating the mobility vs. particle size curve, yet fails to make the calculated curve approach sufficiently close to the experimental one. Calculation of the mobility curve by the Overbeek equation offers no improvement.

Almost similar conclusions can be made in the case of the most dilute solution systems for all three counterions studied. It is found that even for the assumption of the highest useful ζ values [$\zeta = 156$ mV for K^+; $\zeta = 78.14$ mV for Ba^{+2}; and $\zeta = 52.6$ mV for La^{+3}] the calculated mobility curves only just about approach the experimental ones in the case of K^+ and La^{+3}. In the case of Ba^{+2} however (results not shown), the calculated mobility curve is still distinctly below the observed one.

(ii) At the higher two concentrations in case of each counterion (results shown for highest concentration Ba^{+2} only, figure 6(c)) the W.L.O. relaxation curves are generally more successful, for quite legitimate assumed values of ζ.

(iii) In the case of castor-oil droplets, the calculated W.L.O. curve is in some cases very strikingly successful in reproducing the experimental results e.g., the case of the most dilute KCl solution (fig. 7b). More

---

[2]) The curves for the distilled water system in these two cases (figs. 2 and 4 respectively) are placed entirely in keeping with the two different trends mentioned in the two cases, i.e., topmost in figure 2 (dimethylaniline) and bottommost in figure 4 (castor-oil).

generally however the experimental mobilities show a steeper rise with particle size than can be reproduced by the W. L. O. equation even with the best choice of the ζ value, with the result that often the theoretical W. L. O. curve cuts horizontally across the course of the rising experimental points. This is more marked in the case of the higher valent counterions (Ba^{+2}; fig. 7c; La^{+3}, fig. 7d).

(B) *Liquid drop correction treatment:*

Among the two liquid drop electrophoresis equations used the simpler equation based on the assumption of interfacial layer charge distribution is seen to be definitely less successful; in all cases it predicts only a very slow increase of mobility with particle size (figs. 6a, b, c, d; 7a, b, c, d). The equation based on the model of a diffuse double layer type of charge distribution inside the liquid droplets also, is hardly more successful. Generally, it mostly predicts a far too steep rise of mobility with particle size (figs. 6, 7). After the steep initial rise, the mobility curves run almost flat, or else show a dip. Moreover the ζ values required are generally almost always small; as remarked earlier (1, 16), this is a serious limitation for this equation.

On the whole, the relaxation correction equation is definitely more successful than the liquid drop electrophoresis equations.

Conclusion

We have been considering systems in which the ionic composition and concentration of the medium has not been allowed to change. (In addition to the data presented here, see the calculations based on Mooney's and Stackelberg et al's. data). It was therefore plausible to assume that the surface characteristics of the particle, e.g., its electrostatic potential difference with respect to the bulk solution, or its surface charge density, also remain unaltered. In fact, the condition of constant ionic composition and concentration had deliberately been imposed in order to secure the above simplifying situation, namely constant ζ-potential or constant surface charge density.

However, all the above systems studied here, as well as those studies earlier, lead to an identical result, namely the ζ-potential changes (generally increase) with increasing particle size even at constant electrolyte concentration. This result is in contradiction with our basic assumption. As the computer calculation method gives, so to say, a rigorous solution of the equations of the theory, therefore, the above contradiction cannot be attributed to possible approximations in the method of calculation. Hence, the said contradiction must be attributed to either or both of the following reasons: (i) The invalidity of some of the other basic assumptions made in the theories; (ii) A genuine change of the ζ-potential with particle size, even in media of constant electrolyte concentration.

The effect of particle curvature on ζ-potential has been investigated. The next logical step would be a more precise theoretical formulation of the actual nature of this variation with particle radius. Once this is known, the nature of variation of electrophoretic mobility with particle size could then be formulated more correctly, and subsequently put to experimental verification.

Further, we have relied heavily on experimental mobility measurements using droplets of different liquids. Though this has the advantage of producing suspensions of almost perfectly spherical particles; still, the condition of particle rigidity is thereby lost, and the application of rigid-particle relaxation correction equations becomes questionable. The rather meagre success of the Booth liquid drop electrophoresis equations raises doubts as to how far the droplets formed from medium or high viscosity liquids do actually behave as fluid particles; it would appear that more often than not the tiny fluid motions within the drop are arrested by trace contaminations present in the solution.

Summary

Electrophoretic mobility measurements have been carried out using droplets of four different oil-in-water emulsions in presence of electrolytes having different valent counterions. For each electrolyte three different (low) concentrations were used; at each concentration mobilities were measured for particles of varying sizes. In this manner mobilities for different values of $\varkappa a$, though presumably at constant values of ζ, were obtained.

The ζ potentials were calculated by using the Wiersema-Loeb-Overbeek (W. L. O.) computational method, or failing this, the Overbeek analytical approximation method. Assuming suitable constant values of ζ the particle mobilities were then calculated by W.L.O. method. Also, the particles being fluid in nature, the effect, if any, of the viscosity of the fluid of the particle (rather than of the second order electrical effects such as relaxation correction) on its electrophoretic mobility was also investigated. The Booth liquid drop electrophoresis equations were used to calculate the particle mobilities, assuming suitable constant values of (i) $\varkappa/\varkappa'$ and ζ, or (ii) ζ only.

In every case, the measured electrophoretic mobilities have been found to increase with increasing particle size. The effect of electrolytes is, in general, to maintain the same trend. Again, the calculated ζ-potentials have been found not to remain constant in solutions of constant electrolyte concentration and composition, but show a steady increase with increasing particle size; this is true for both the distilled water as also the electrolyte solution systems.

The calculation of electrophoretic mobility, using the W.L.O. relaxation correction method, on the assumption of a suitable average (constant) value of ζ for each system shows that the calculated mobility values agree generally well with the observed values. Also, among the two liquid drop electrophoresis equations used, the simpler equation based on the assumption of interfacial layer charge distribution is found to be definitely less successful in all cases; it predicts only very slight increase of mobility with particle size. The equation based on the model of a diffuse double layer type of charge distribution inside the liquid droplet is somewhat more successful.

Zusammenfassung

Es wird die elektrophoretische Beweglichkeit der Tröpfchen von Öl/Wasser-Emulsionen untersucht. Die experimentellen Daten werden nach dem Verfahren von Wiersema-Loeb-Overbeek (W.L.O.) und Booth ausgewertet. In den vier untersuchten Fällen steigt sie mit dem Tröpfchenradius an. Das gilt sowohl für Emulsionen in reinem Wasser als auch in Elektrolytlösungen. Einige mögliche Gründe für dieses Verhalten werden diskutiert.

References

1) *Sengupta, M., D. N. Biswas*, J. Colloid Interface Sci. **29**, 536 (1969); **30**, 275 (1969).

2) *Sengupta, M., S. S. Jana*, Indian J. Chem. **10**, 509 (1972).

3) *Jana, S. S., D. N. Biswas, M. Sengupta*, J. Indian Chem. Soc. **47**, 527 (1970).

4) *Overbeek, J. Th. G.*, Kolloid Chem. Beih. **54**, 287 (1943).

5) *Booth, F.*, Proc. Roy. Soc. **203**, 514 (1950).

6(a) *Wiersema, P. H., A. L. Loeb, J. Th. G. Overbeek*, J. Coll. Interface Sci. **22**, 78 (1966).

6(b) *Wiersema, P. H.*, Ph.D. Thesis (Utrecht 1964).

7) *Mooney, M.*, Phys. Rev. **23**, 396 (1924).

8) *Stackelberg, M. V., H. Heindze, F. Wilke, R. Doppelfeld*, Z. Elektrochem. **61**, 781 (1957).

9) *Kemp, I.*, Trans. Far. Soc. **31**, 1347 (1935).

10) *Shaw, J. N., R. H. Ottewill*, Nature **208**, 681 (1965).

11) *Vogei, A. I.*, Quantitative Inorganic Analysis, 3rd. edn. (1962).

12) *Bull, H. B.*, J. Am. Chem. Soc. **80**, 1901 (1958).

13) *Ives, D. J. G., G. J. Janz*, Reference Electrodes, (New York 1961).

14) *Kruyt, H. R.*, ed., Colloid Science, vol. 1 (Amsterdam 1952).

15) *Sengupta, M., S. S. Jana, D. N. Biswas*, communicated for publication.

16(a) *Jana, S. S.*, Ph. D. Thesis, Calcutta University (1973).

16(b) *Biswas, D. N.*, Ph. D. Thesis, Calcutta University (1974).

17(a) *Robinson, R. A., R. H. Stokes*, Electrolyte Solutions, p. 465 (London 1959).

17(b) *Kaye, G. W. C., Laby, T. H.*, Tables of Physical and Chemical Constants, 12th edn., p. 36 (1959).

18) *Hunter, R. J.*, J. Phys. Chem. **66**, 1367 (1962).

19) *Sengupta, M.*, Indian J. Chem. **5**, 648 (1967).

20) *Biswas, D. N., M. Sengupta*, J. Indian Chem. Soc. **44**, 1042 (1967).

Authors' address:

S. S. Jana, M. Sengupta, and *D. N. Biswas*
Department of Chemistry,
University College of Science
Calcutta 700009 (India)

Progr. Colloid & Polymer Sci. **63**, 87–89 (1978)

ISSN 0340-255 X

Instruments Research and Development Establishment, Dehra Dun (U.P.), (India)

The influence of dispersed phase volume and particle size distribution on the rheology of oil-in-oil emulsions

M. K. Sharma

With 3 figures and 1 table

(Received December 13, 1976)

It has been shown earlier (1), that in addition to normal oil-in-water and water-in-oil emulsions, a new third type of oil-in-oil emulsion can be prepared with two nonaqueous immiscible liquids as continuous and disperse phases and surfactant as emulsifying agent.

Many factors influence the rheological properties of emulsions (2). The knowledge of the effects of dispersed phase volume and particle size distribution on viscosity is most important to study the rheological behaviour of an emulsion.

Following *Sweeney* and *Geckler* (3) and *Saunders* (4), the influence of dispersed phase volume and particle size on relative viscosity (ηrel) can be defined by

$$\eta_{\mathrm{rel}} = \exp\left(\frac{a\phi}{1 - k\phi}\right) \qquad [1]$$

where a is a constant, ϕ is the volume fraction of the dispersed phase and k is a hydrodynamic interaction coefficient which depends on mean particle size.

Eq. [1] can be rewritten as

$$\frac{\phi}{\log \eta_{\mathrm{rel}}} = -\frac{2{\cdot}303\, k\phi}{a} + \frac{2{\cdot}303}{a} \qquad [2]$$

so that both a and k can be derived from a linear plot of ϕ/log ηrel against ϕ.

The mean particle size (Dm) of globules in an emulsion is given by (5).

$$Dm = \left\{\frac{\sum nD^3}{\sum n}\right\}^{\frac{1}{3}} \qquad [3]$$

where n is the number of droplets and D is the droplet diameter.

In the present investigations attempts are being made to study the effects of dispersed phase volume and particle size distribution on the viscosity of oil-in-oil emulsions with monochlorobenzene and ethylene glycol as two nonaqueous phases and polyoxyethylene sorbitan monooleate as emulsifier.

Experimental

Emulsion preparation. Monochlorobenzene-in-ethylene glycol nonaqueous emulsions were prepared using 2.0% concentration of polyoxyethylene sorbitan monooleate (KL) (w/v % of emulsion). The volume fractions of monochlorobenzene (BDH) ranging from 0.1111 to 0.4286 were employed. The heterogeneous mixtures of monochlorobenzene and ethylene glycol (BDH) containing nonionic surfactant were emulsified with the help of Braun emulsator for final making of the emulsions.

The emulsions are coded in table 1 according to the dispersed phase concentration used. In either series of samples, O/O emulsions were prepared with one of five different dispersed phase concentrations. Portions of each of these nonaqueous emulsions were then diluted with the ethylene glycol phase containing the relevant concentration of nonionic surfactant to give still lower dispersed phase concentrations. Analysis of particle size distribution as described below, confirmed that this dilution procedure did not change the original particle size distribution.

Particle size analysis. Particle size distribution of O/O emulsions was determined by photomicrographic method (6), by taking photomicrographs of microscopic slides of emulsions on ORWO, NP 27, 400 ASA cut films with the help of Carl Zeiss Jena microscope equipped with an attachment camera using a 25X projection system and a 40X objective, as described in previous work (1).

Viscosity measurements. A Weissenberg rheogoniometer model R.16 (Sangamo Weston, Bognor Regis, England) was used for viscosity determinations. A high rate of shear (1467.6 sec^{-1}) was employed.

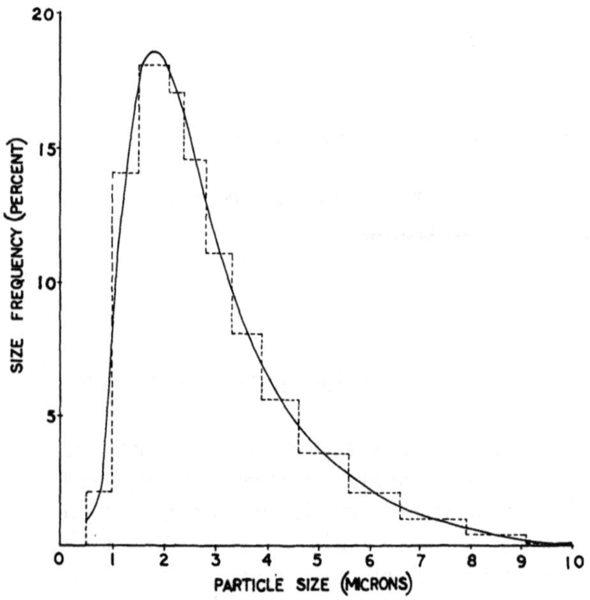

Fig. 1. Particle size distribution in a typical nonaqueous emulsion

Cone diameter and cone angle were 7.5 cm and 1°32′ respectively. A constant temperature of 30.0° ±0.1 °C was maintained by means of a water jacket, and all samples were left between the cone and plate for one hour to attain this temperature before viscosity measurements were carried out.

Results

Particle size distribution data of a typical nonaqueous emulsion '*A*' prepared with 0.1111 dispersed phase concentration and 2.0% emulsifier concentration is given in figure 1. The O/O emulsion contains low percentage of particles less than 1.0 μm. The values of mean particle size (*Dm*) calculated from eq. [3] for the emulsions series *A* to *E* are given in table 1. The viscosity data for the original and diluted nonaqueous emulsions are also shown in table 1.

Plot of the reduced viscosity $\eta sp/\phi$ against ϕ in figure 2, where ηsp is the specific increase in the viscosity, indicates a nonlinear dependence

Table 1. Particle size and viscosity data of nonaqueous emulsions

Emulsion No.	ϕ	ηrel	$\eta sp/\phi$	$\phi/\log \eta$rel	*a*	*k*	*Dm*
A	.1111	1.500	4.500	.6309			
A1	.0870	1.312	3.586	.7379			
A2	.0638	1.194	3.041	.8286	2.193	3.580	3.470
A3	.0417	1.112	2.686	.9046			
A4	.0204	1.050	2.451	.9623			
B	.1765	1.819	4.640	.6794			
B1	.1364	1.524	3.842	.7454			
B2	.0989	1.316	3.195	.8290	2.152	2.164	3.561
B3	.0638	1.172	2.696	.9260			
B4	.0309	1.073	2.362	1.0098			
C	.2500	2.298	5.192	.6918			
C1	.1905	1.772	4.052	.7666			
C2	.1364	1.450	3.299	.8451	2.101	1.557	3.634
C3	.0870	1.238	2.736	.9385			
C4	.0417	1.098	2.350	1.0271			
D	.3333	3.020	6.061	.6944			
D1	.2500	2.118	4.472	.7671			
D2	.1765	1.602	3.411	.8622	2.084	1.188	3.701
D3	.1111	1.304	2.736	.9636			
D4	.0526	1.124	2.357	1.0354			
E	.4286	4.047	7.109	.7060			
E1	.3158	2.516	4.801	.7881			
E2	.2195	1.768	3.499	.8869	2.038	0.964	3.728
E3	.1364	1.375	2.749	.9863			
E4	.0638	1.145	2.273	1.0850			

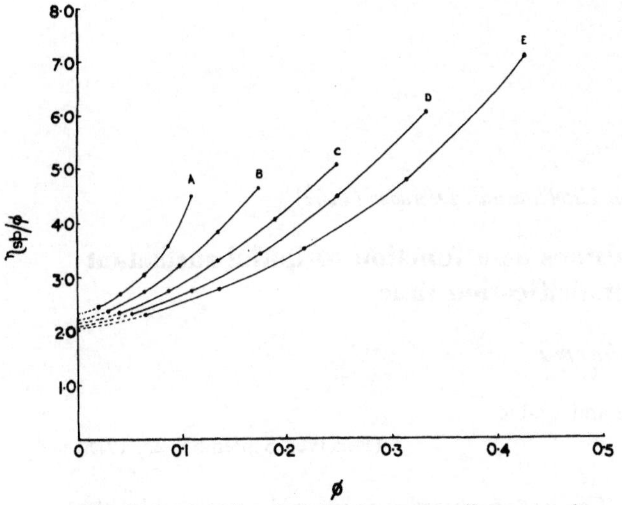

Fig. 2. Reduced viscosity versus disperse phase concentration for nonaqueous emulsions

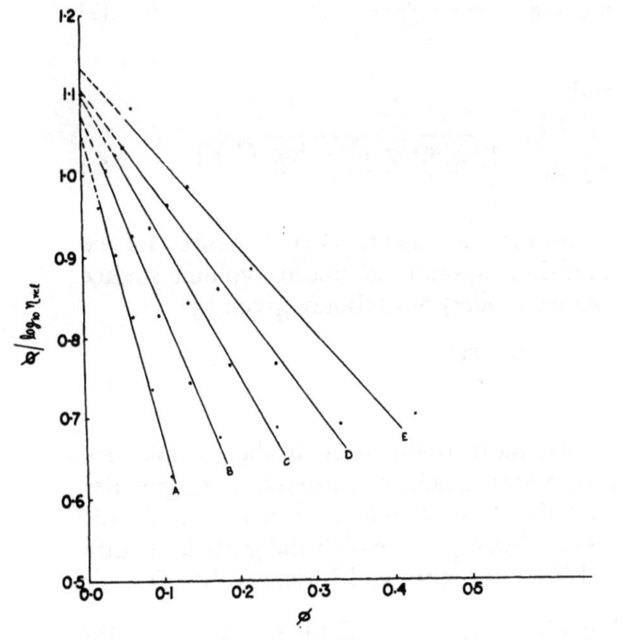

Fig. 3. Plot of ϕ/log ηrel versus ϕ for nonaqueous emulsions

over the dispersed phase concentration range from 0.1111 to 0.4286 and also shows quite clearly that mean particle size has a definite effect on the viscosity data. Of course, the particle size distribution will have an effect also.

In figure 3, is shown a plot of ϕ/log ηrel against ϕ for nonaqueous emulsions series A to E with different mean particle size indicating a reasonably good linear relationship for each particle size. Relevant values of a and k obtained from such plots are given in table 1. The values of constant a for the nonaqueous emulsions have been found to be lower than 2.5 which shows that the particles do not behave as rigid spheres and particle aggregates are present in the emulsions. The values of k indicate a dependence of this interaction coefficient on mean particle size, decreasing with increasing particle size.

Thus the relative viscosity of oil-in-oil emulsions increases with an increase in dispersed phase concentration from 0.1111 to 0.4286, with a corresponding increase in mean particle size from 3.470 μm to 3.728 μm, while the emulsifier concentration is kept constant.

Summary

The effects of dispersed phase volume and particle size distribution on the rheology of oil-in-oil emulsions (nonaqueous systems) prepared with monochlorobenzene and ethylene glycol as disperse and continuous phases and polyoxyethylene sorbitan monooleate as emulsifier, have been studied. Experimental viscosity data obtained for nonaqueous emulsions appeared to fit the general equation $\eta\text{rel} = \exp\ (a\phi/1 - k\phi)$. The values of constant a have been found to be below the value of 2.5 for solid spheres and the hydrodynamic interaction coefficient k decreases with an increase in mean particle size.

Zusammenfassung

Es wurde der Einfluß des Volumens und der Teilchengrößenverteilung der dispersen Phase auf die Rheologie von Öl-in-Öl-Emulsionen untersucht. Die Emulsionen wurden mit Monochlorbenzol und Äthylenglycol als disperser bzw. kontinuierlicher Phase und Polyoxyäthylensorbitanmonooleat als Emulgator hergestellt.

References

1) *Sharma, M. K.*, J. Colloid Interface Sci. **53**, 340 (1975).

2) *Sherman, P.*, Emulsion Science, p. 285 (London 1969).

3) *Sweeney, R. H., R. D. Geckler*, J. Appl. Phys. **25**, 1135 (1954).

4) *Saunders, F. L.*, J. Colloid Sci. **16**, 13 (1961).

5) *Becher, P.*, Emulsions: Theory and Practice, p. 54 (New York 1966).

6) *Shillaber, C. P.*, Photomicrography in Theory and Practice, p. 41 (New York 1949).

Author's address:

M. K. Sharma
Instruments Research and Development Establishment
Dehra Dun (U. P.)
India

Progr. Colloid & Polymer Sci. **63**, 90–95 (1978)

ISSN 0340-255 X

Instruments Research and Development Establishment, Dehradun (India)

Globule size distribution of oil-in-oil emulsions as a function of initial surfactant concentration and emulsification time

M. K. Sharma

With 6 figures and 1 table

(Received September 25, 1975)

Introduction

The formation, properties, stability, and instability of oil-in-water and water-in-oil emulsions have been studied extensively (1, 2). It is really surprising that proper attention has not been paid so far to the chemistry of a new possible third type of emulsion, which may be called as oil-in-oil (O/O) emulsion, i.e., nonaqueous system formed by the intimate dispersion of an immiscible liquid in another (both nonaqueous), in the form of globules, in presence of stabilizing agent.

In the present investigations, globule size distribution of oil-in-oil emulsions with benzene and ethylene glycol as two nonaqueous phases and sodium dioctylsulphosuccinate (anionic surfactant) as emulsifier, has been studied and an effort has been made to find out the effects of initial surfactant concentration and time of emulsification on globule size of oil-in-oil emulsions.

In dispersed systems, according to Gaussian or normal distribution law, the logarithmic size frequency distribution curve is defined by (3)

$$G(D) = \frac{\sum n}{2.303 \log \sigma_g \sqrt{2\Pi}} \exp\left[-\frac{(\log D - \log D_g)^2}{2 \log^2 \sigma_g}\right] \quad [1]$$

where $G(D)$ is the probability or frequency of observation of droplet diameter D, n is the number of droplets within size interval ΔD, Dg is the geometric mean diameter of globules and σ_g is the geometric standard deviation. The last two parameters are determined by the expressions

$$\log Dg = \frac{\sum n \log D}{\sum n} \quad [2]$$

and

$$\log \sigma_g = \sqrt{\frac{\sum[n(\log D - \log Dg)^2]}{\sum n}} \quad [3]$$

Another parameter, generally desired for the emulsion system is mean volume-surface diameter (Dvs) of globules, given by

$$Dvs = \frac{\sum n\, D^3}{\sum n\, D^2} \quad [4]$$

One more useful term is the globule size Dmf about which are clustered those globules with the greatest frequency of occurrence or, more precisely, for which the globule density $n/\Delta D$ is maximum and is defined by (4)

$$\log Dmf = \log Dg - 2.303 \log {}^2\sigma_g \quad [5]$$

Experimental

Materials: Ethylene glycol and benzene employed were of BDH AR grade. Sodium dioctylsulphosuccinate (Manoxol OT) was BDH England product. These chemicals were used without further purification.

Formation of Emulsions: The oil-in-oil emulsions were prepared with equal volumes of benzene and ethylene glycol as two immiscible fluid phases and different initial concentrations of Manoxol OT (W/V% of ethylene glycol) as emulsifying agents. Surfactant was dissolved in ethylene glycol before emulsification. Benzene was added slowly to the respective solutions of emulsifier in ethylene glycol with constant stirring. The heterogeneous mixtures thus made were emulsified with the help of Braun emulsator at a constant speed, for final making of the emulsions. The starting temperature of emulsification was kept at 30.0° ± 1.0 °C and the temperature was never allowed to increase above 40 °C during emulsification. Small quantities of samples of emulsions were withdrawn from time to time during emulsification for globule size determinations, in order to find out the time to reach a steady state in globule size.

The conditions described above were reproduced exactly in the preparation of each emulsion. The sensitivity of the globule size of O/O emulsions to changes in the numerous variables is not yet known, and for this reason all variables except surfactant concentration and emulsification time were constant to the degree indicated in order that variations in globule size could be definitely attributed to the concentration of surfactant and time of emulsification.

Immediately after preparation, the emulsion type was determined by dye solubility method (5) and all the emulsions were found to be benzene in ethylene glycol type.

Globule Size Analysis: Particle size distribution of nonaqueous emulsions were determined by photomicrographic method (6). Samples of emulsions were diluted to 1% internal phase with ethylene glycol and mounted on microscopic slides with cover glasses. Photomicrographs were then taken on ORWO, NP 27, 400 ASA cut films with the help of Carl Zeiss Jena microscope equipped with attachment camera using 25 × projection system and 40 × objective. An exposure time of 0.02 second was used, by adjusting the illumination, so that motion of the suspended emulsion droplets due to Brownian movement was effectively "stopped". At least three different areas of the sample slide were photographed. After the photomicrographs were taken, a stage micrometer, with a ruling 1 mm long and divided into 100 equal parts, was placed on the microscope stage and all the camera and microscope adjustments were made exactly same as they were, when the photograph was taken. With the lines of the stage micrometer carefully focussed on the ground glass of the camera, the image was measured with an ordinary millimeter scale. The magnification factor was calculated by the following formula (7)

$$\text{Magnification} = \frac{\text{Distance in the image field}}{\text{Equivalent distance in the object field.}} \qquad [6]$$

The processed negatives were placed on a milk glass plate illuminated from the bottom and individual diameters of 400 to 500 droplets were determined by dividing the observed diameter with the magnification factor.

In order to see the reproducibility of the results, duplicate emulsions prepared under identical conditions were investigated. Average precision of the mean of duplicate determinations was ±2% of the reported values.

Table 1. Globule Size Distribution Data of a Typical Nonaqueous Emulsion
$Dg = 2.853\ \mu$; $\sigma_g = 1.607\ \mu$; $Dvs = 4.585\ \mu$; $Dmf = 2.278\ \mu$

(1) Size range (μ)	(2) Mid size D (μ)	(3) n	(4) $\frac{n}{\Delta D}$	(5) $\frac{nD}{\Delta D}$	(6) $G(D)$	(7) $n \log D$	(8) nD^2	(9) nD^3	(10) Cumulative n	(11) Cumulative percent ncp (%)	(12) % Frequency $\frac{n}{\sum n}$ (%)
0 –0.5	.25	0	0	0	0	0	0	0	0	0	0
0.5–1.0	.75	8	16	12.0	7.72	−1.00	4.50	3.38	8	1.65	1.65
1.0–1.5	1.25	36	72	90.0	89.80	3.49	56.25	70.31	44	9.07	7.42
1.5–2.0	1.75	67	134	234.5	239.74	16.28	205.19	359.08	111	22.89	13.81
2.0–2.5	2.25	80	160	360.0	359.76	28.18	405.00	911.25	191	39.38	16.49
2.5–3.0	2.75	72	144	396.0	406.57	31.63	544.50	1497.38	263	54.23	14.85
3.0–3.5	3.25	59	118	383.5	392.67	30.20	623.19	2025.36	322	66.39	12.16
3.5–4.0	3.75	45	90	337.5	345.40	25.83	632.81	2373.05	367	75.67	9.28
4.0–4.5	4.25	33	66	280.5	286.48	20.74	596.06	2533.27	400	82.47	6.80
4.5–5.0	4.75	23	46	218.5	228.81	15.56	518.94	2464.95	423	87.22	4.74
5.0–6.0	5.5	30	30	165.0	156.51	22.21	907.50	4991.25	453	93.40	6.19
6.0–7.0	6.5	15	15	97.5	90.37	12.19	633.75	4119.38	468	96.49	3.09
7.0–8.0	7.5	8	8	60.0	51.14	7.00	450.00	3375.00	476	98.14	1.65
8.0–9.0	8.5	3	3	25.5	28.86	2.79	216.75	1842.38	479	98.76	.62
>9	–	6	–	–	–	~5.73	–	–	485	100.00	1.24
Total	–	485	902	2660.5	2683.83	220.83	5794.44	26566.04	–	–	99.99

Results and discussion

Size frequency distribution data of a typical nonaqueous emulsion prepared with equal volumes of benzene and ethylene glycol and 2.5% initial surfactant concentration, is summarised in table 1. The time of emulsification was exactly 5 min. insuring that the emulsion reached a steady state in globule size distribution. In the first three columns of table 1 are tabulated the essential data of the emulsion. The number of globules n falling within a given size interval ΔD is shown in column 3, the midpoint of the interval being D. The values of $G(D)$, Dg, σ_g, Dvs, and Dmf of the emulsion were calculated from Eqs. [1] through [5]. The parameters of the particular system are $Dg = 2.853$ μ, $\sigma_g = 1.607$ μ, $Dvs = 4.585$ μ, and $Dmf = 2.278$ μ.

The size or size-frequency distribution of a particular nonaqueous emulsion can be explained with a knowledge of the globule density, i.e., the extent to which globules are grouped by count about some given diameter. The most convenient expression for globule density is globules per unit size interval, $n/\Delta D$. By the simple expedient of division by ΔD (column 4 in tab. 1), all n values are normalized to an interval of 1 micron. The plot of $n/\Delta D$ vs. the mid-interval diameter D is shown in figure 1. Such a procedure closely approximates the number of droplets which would have been observed if equal size interval of 1 micron were originally taken and yields a curve, true representative of the system. The curve reaches its peak value at the globule diameter of greatest frequency of occurrence Dmf.

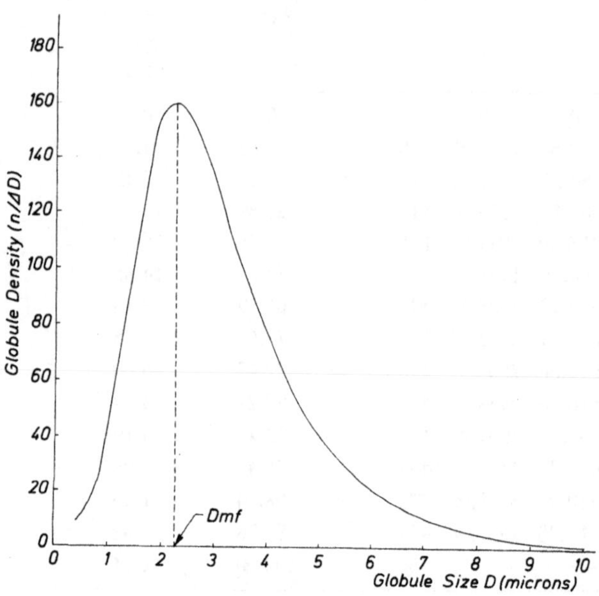

Fig. 1. Globule density vs. size

The physical significance of frequency function $G(D)$ is that it is numerically equal to $nD/\Delta D$, i.e., a quantity obtained by weighing the globule density $n/\Delta D$ by the midinterval size D. Values of $nD/\Delta D$ appear in column 5 of table 1; column 6 gives the values for $G(D)$ calculated from Eq. [1]. Each $nD/\Delta D$ value is not in close agreement with its corresponding $G(D)$ value. The reason they are not identical is that the equation yields values based upon a perfectly smooth curve representing an idealised model, whereas our calculations are based upon practical data: we are obliged to make our observations over finite intervals ΔD, counting globules only to the nearest whole number. For the same reason, the calculation of Dg, σ_g, Dvs, and Dmf from Eqs. [2] through [5], utilizing the data of table 1, yields values almost, but not quite, identical to the exact values.

If we plot globule density $n/\Delta D$ against logarithm of the size interval D, the curve becomes symmetrical as shown by the dotted curve of, figure 2, and its peak value remains the same as earlier, i.e., at the globule diameter of greatest frequency of occurrence Dmf. If, however, we take each point on this curve and weigh it according to the corresponding particle size D, the resulting values of $nD/\Delta D$ yield the solid curve of figure 2, which is also symmetrical but reaches its peak at the number median diameter Dg of 2.853 μ. Since $nD/\Delta D$ and $G(D)$ are numerically equal, the solid curve is effectively a plot of $G(D)$ vs. log D. The latter curve graphically illustrates that the droplets are symmetrically grouped in some coherent manner with respect to the logarithms of their diameters, the axis of symmetry being the logarithm of the number geometric diameter Dg. The curves of figure 2 thus illustrate a graphical method for determining Dg and Dmf of nonaqueous emulsions.

A more expedient graphical method to determine the values of Dg and σ_g is to make use of logarithmic probability paper, plotting the percent ncp of all globules less than a stated size vs. the logarithm of the stated size

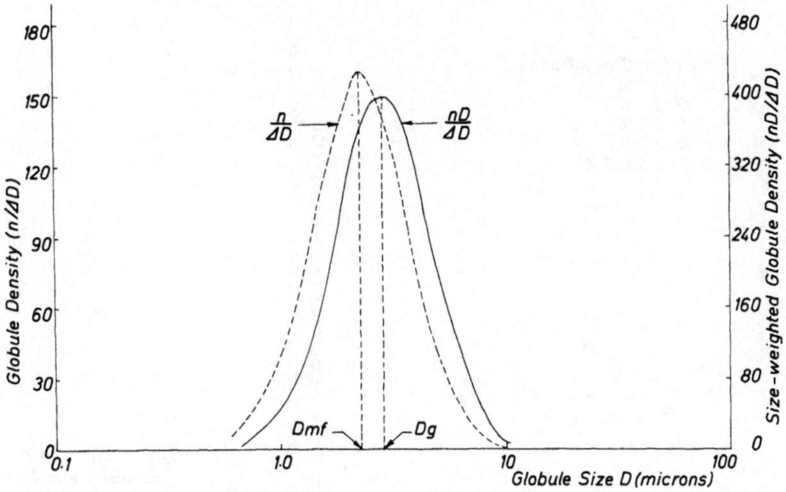

Fig. 2. Globule density and size weighed globule density vs. size

as shown in figure 3. It is important to note that this cumulative percent *ncp* should be plotted against the upper limit of the interval ΔD, not the mid-value. It can readily be seen that if *n* globules appear in the interval ΔD, these and all smaller globules are, by definition, smaller than the upper limit of ΔD, rather than its mid-value.

From the log-probability curve we can at once determine the two parameters Dg and σ_g which rigidly define the emulsion system. By the definition of Dg, its value is found at the 50% mark. And it can be shown that σ_g is given by

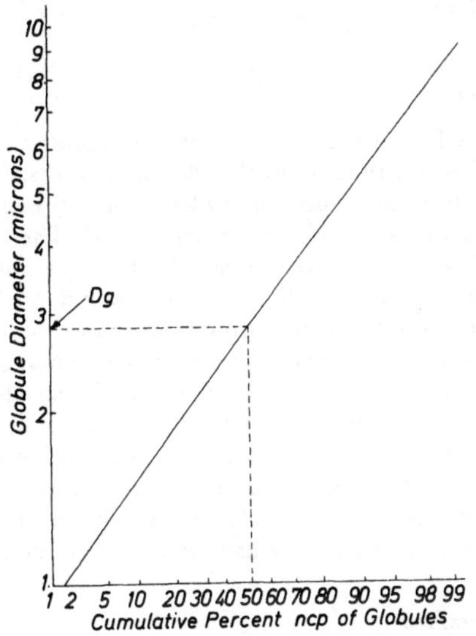

Fig. 3. Cumulative per cent of globules less than a stated size vs. the stated size

$$\sigma_g = \frac{84.13\,\%\ \text{size}}{50\,\%\ \text{size}} = \frac{50\,\%\ \text{size}}{15.87\,\%\ \text{size}}\,. \qquad [7]$$

The influence of emulsification time to reach a steady state in globule size of oil-in-oil emulsions prepared with six different initial surfactant concentration is shown in figure 4. An emulsion prepared with an initial surfactant concentration of 1% came to a steady state in globule size, expressed in mean volume-surface diameter in approximately 120 seconds. With initial surfactant concentration from 2 to 4%, the emulsions reached steady state in globule size distribution in approximately 180 to 300 seconds, respectively, thereafter with an increase in surfactant concentration, the emulsification time required to reach steady state globule size remains constant. Thus the time required to reach a steady state globule diameter evidently goes up with the initial surfactant concentration used to stabilize the emulsion, upto an optimum concentration of emulsifying agent (fig. 5).

The increase in the emulsification time to reach a steady state in globule size of non-aqueous emulsions as the initial surfactant concentration is increased may well be related to the increase in the viscosity of dispersion medium, which becomes constant when an optimum viscosity of emulsions is achieved. This may be correlated with the rheological properties of aqueous emulsions, where an increase in the initial surfactant concentration results in an increase in the viscosity of the dispersion medium resulting in corresponding

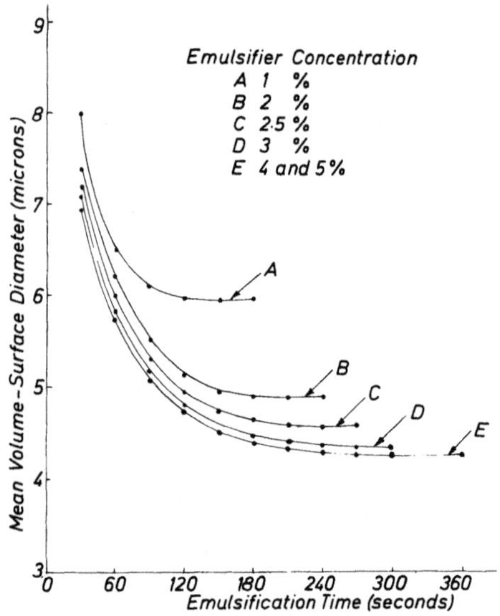

Fig. 4. Emulsification time vs. mean volume – surface diameter

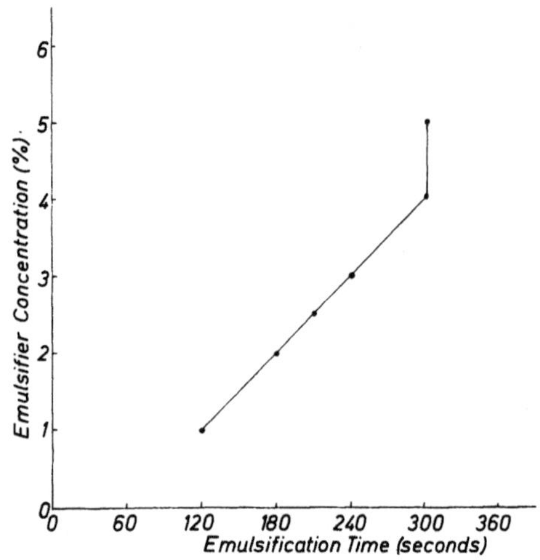

Fig. 5. Emulsification time vs. emulsifier concentration

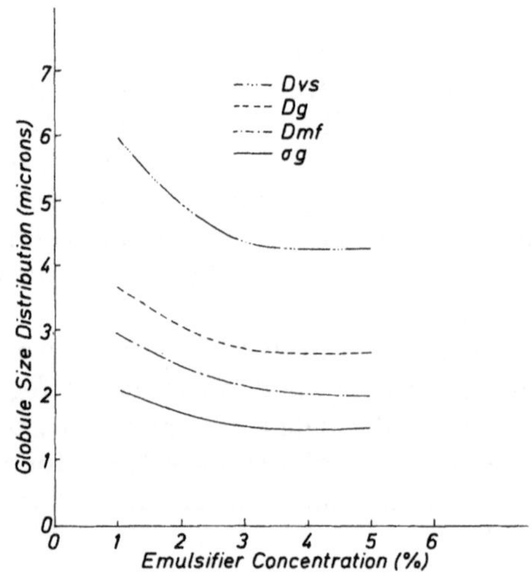

Fig. 6. Emulsifier concentration vs. globule size distribution

decrease in the diffusion rate (8) of the surfactant molecules to the surface of dispersed droplets; thus a longer time would be required for sufficient surfactant to be adsorbed at the interface to stabilize the dispersed droplets, against coalescence (9).

The effect of initial emulsifier concentration on *Dvs*, *Dg*, *Dmf*, and σ_g of nonaqueous emulsions prepared with equal volumes of benzene and ethylene glycol is shown in figure 6. The emulsification time for all these emulsions was kept exactly 5 min., insuring that the emulsions reached a steady state in globule size distribution. During emulsification the temperature rise was also same for these emulsions insuring that the viscosity and density changes would be similar. The increase of Manoxol OT concentration from 1 to 3%, slightly reduces the values of *Dvs*, *Dg*, *Dmf*, and σ_g and further increase of the concentration upto 5% causes almost negligible reduction of the values. Thus, as the emulsifier concentration is increased, the nonaqueous emulsions are easily formed, the globules become smaller in size and the emulsions are stable to some extent. But the increase of emulsifier concentration beyond an optimum value neither improves the stability nor gives finer droplets. Thus the initial surfactant concentration is not a major factor and insignificant to influence the steady state globule size distribution of nonaqueous emulsions.

Summary

The effects of initial emulsifier concentration and emulsification time upon the globule size distribution of new type of emulsions, called oil-in-oil emulsions (nonaqueous systems) have been studied. The time of emulsification to reach a steady state globule size increases linearly with increasing initial surfactant concentration, upto an optimum emulsifier concentration, thereafter it remains constant. The initial emulsifier concentration has little effect on the steady state globule size distribution of nonaqueous emulsions. The values of frequency function, geometric standard deviation, geometric mean diameter, mean volume-surface diameter, and globule diameter of most frequent occurrence, of emulsions have been determined.

Zusammenfassung

Der Einfluß der Emulgatorkonzentration und der Emulgierungszeit auf die Korngrößenverteilung neuer

Emulsionstypen (Öl-in-Öl-Emulsionen) wurden untersucht. Die Emulgierungszeit bis zur Erreichung einer konstanten Kügelchengröße nimmt linear mit wachsender Anfangskonzentration des oberflächenaktiven Agens (Emulgators) zu, bis eine optimale Emulgatorkonzentration erreicht ist; danach bleibt die Emulgierungszeit konstant. In nichtwässerigen Emulsionen hat die anfängliche Emulgatorkonzentration wenig Einfluß auf das Spektrum der Kügelchengrößen im stationären Zustand. Die Verteilungsfunktion, die Standardabweichung, der geometrisch mittlere Durchmesser und der am häufigsten vorkommende Durchmesser sowie das mittlere Volumen-Oberflächenverhältnis wurden bestimmt.

References

1) *Clayton, W.*, Theory of Emulsions and their Technical Treatment (London 1954).

2) *Becher, P.*, Emulsions: Theory and Practice, 2nd ed., (New York 1965).

3) *Dalla Valla, J. M.*, in: Micromeritics, 2nd ed., (New York 1948).

4) *Aitchison, J., J. A. C. Brown*, The Lognormal Distribution (Cambridge 1957).

5) *Hauser, E. A., J. E. Lynn*, Experiments in Colloid Chemistry, p. 129 (New York 1940).

6) *Sherman, P.*, Emulsion Science, p. 153 (London 1969).

7) *Shillaber, C. P.*, Photomicrography in Theory and Practice, p. 41 (New York 1949).

8) *Clifford, J., B. A. Pethica*, Trans. Faraday Soc. **60**, 216 (1964).

9) *Rajagopal, E. S.*, Kolloid-Z. **162**, 7 (1959).

Author's address:
M. K. Sharma
Instruments Research and Development Establishment
Dehradun (UP) (India)

Progr. Colloid & Polymer Sci. **63**, 96–98 (1978)

ISSN 0340-255 X

Politechnika Poznańska, Institute of Chemical Technology, Poznań (Poland)

The new method of hydrophobicity index calculation

J. Szymanowski

With 2 figures and 1 table

(Received December 16, 1976)

Introduction

Recently the concept of the effective chain length (n_{eff}), taking into account the influence of the structure of surface active agents on surface phenomena and bulk properties, was introduced (1–4). Several different methods to determine the effective chain length were also described (5).

On the basis of the determined values of the effective chain length of hydrophobe hydrophobicity index (*HI*), defined as a ratio of the effective number of methylene groups in a hydrophobe chain to the actual number in it (n), can be calculated

$$HI = \frac{n_{eff}}{n}.$$

This concept of n_{eff} and *HI* was used to study the behaviour of ionic surfactants in water solutions and to describe the correlations between CMC and other surfactants properties.

The determination of n_{eff} or *HI* requires many accurate, tedious and time-consuming although not difficult experiments. It was decided therefore to look for some calculation methods which could be used to extrapolate and interpolate n_{eff} or *HI* values determined for two members of this homologous series. Such a method is described in this paper. The values of hydrophobicity index cited by *I. J. Lin* and *L. Marszall* (4) were used.

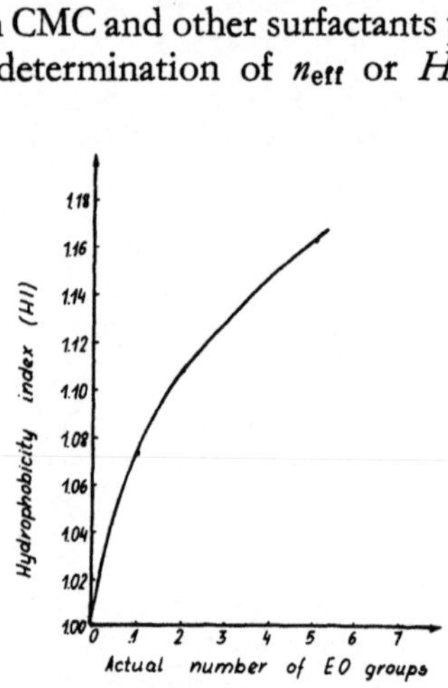

Fig. 1. The relationship between hydrophobicity index and actual number of oxyethylene groups in polyoxyethylene sodium dodecyl sulfates

Problem formulation and discussion

The homologous series of polyoxyethylene sodium dodecyl sulfates was taken as a standard series and then the relationships between

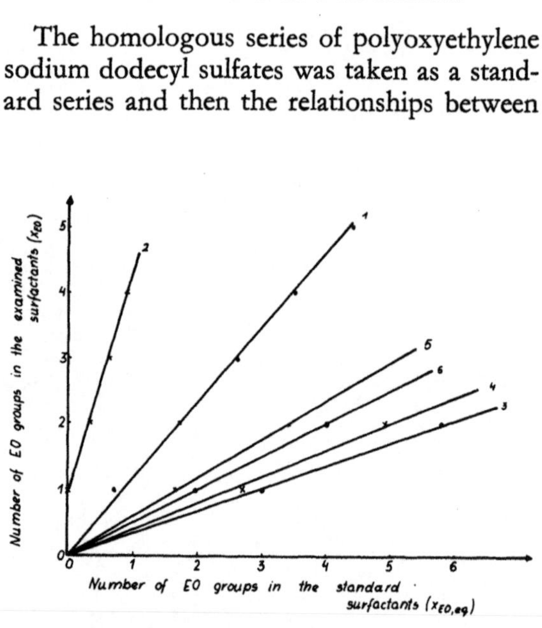

Fig. 2. The relationships between equivalent and actual number of oxyethylene groups / standards – polyoxyethylene sodium dodecyl sulfates
1 – $C_{16}H_{33}(OCH_2CH_2)_xOSO_3Na$
2 – $C_{18}H_{37}(OCH_2CH_2)_xOSO_3Na$
3 – $[C_{10}H_{21}(OCH_2CH_2)_xN(CH_3)_3]Cl$
4 – $[C_{12}H_{25}(OCH_2CH_2)_xN(CH_3)_3]Cl$
5 – $[C_{12}H_{25}(OCH_2CH_2)_xNC_5H_5]Cl$
6 – $[C_{16}H_{33}(OCH_2CH_2)_xNC_5H_5]Cl$

the number of oxyethylene groups in the examined series of surfactants and the standard series for the same values of *HI* were determined. Therefore the relationship between *HI* of polyoxyethylene sodium dodecyl sulfate and the number of EO groups was drawn (fig. 1). Then *HI* values of the examined surfactants were taken into consideration and the equivalent numbers of EO groups in the standard series were read from figure 1.

The linear relationships between the equivalent and actual number of EO groups for the considered series of surfactants, as shown in figure 2, were obtained. These relationships calculated on the basis of the least square method are as follows:

$C_{16}H_{33}(OCH_2CH_2)_xOSO_3Na$

$$X_{EO,eq} = -0.08 + 0.894\,X_{EO}$$

$C_{18}H_{37}(OCH_2CH_2)_xOSO_3Na$

$$X_{EO,eq} = -0.26 + 0.28\,X_{EO}$$

$[C_{10}H_{21}(OCH_2CH_2)_xN(CH_3)_3]Cl$

$$X_{EO,eq} = 0.03 + 2.90\,X_{EO}$$

$[C_{12}H_{25}(OCH_2CH_2)_xN(CH_3)_3]Cl$

$$X_{EO,eq} = 0.08 + 2.45\,X_{EO}$$

$[C_{12}H_{25}(OCH_2CH_2)_xNC_5H_5]Cl$

$$X_{EO,eq} = -0.02 + 1.70\,X_{EO}$$

$[C_{16}H_{33}(OCH_2CH_2)_xNC_5H_5]Cl$

$$X_{EO,eq} = -0.02 + 2.0\,X_{EO}$$

Table 1. Hydrophobicity index of the examined surfactants

No.	Formula	*n*	X_{EO}	$X_{EO,eq}$	*HI* literature	*HI* calculated value	Error absolute	Error relative (%)
1	2	3	4	5	6	7	8	9
1	$C_nH_{2n+1}(OCH_2CH_2)_xOSO_3Na$	12	0	0	1			
			1	1	1.073			
			2	2	1.108			
			3	3	1.119			
			5	5	1.163			
2	$C_nH_{2n+1}(OCH_2CH_2)_xOSO_3Na$	16	0	0	1	1	0	0.00
			1	0.7	1.059	1.058	−0.001	0.09
			2	1.7	1.098	1.098	0.000	0.00
			3	2.6	1.128	1.122	−0.006	0.53
			4	3.5	1.138	1.139	0.001	0.09
			5	4.4	—	1.153	—	—
3	$C_nH_{2n+1}(OCH_2CH_2)_xOSO_3Na$	18	0	0	1	1	0	0.00
			1	0	1.000	1.000	0.000	0.00
			2	0.3	1.029	1.030	0.001	0.08
			3	0.6	1.055	1.053	−0.002	0.19
			4	0.9	1.072	1.069	−0.003	0.28
			5	1.1	—	1.077	—	—
4	$[C_nH_{2n+1}(OCH_2CH_2)_xN(CH_3)_3]Cl$	10	0	0	1	1	0	0.00
			1	3	1.130	1.129	−0.001	0.09
			2	5.8	1.173	1.174	0.001	0.09
5	$[C_nH_{2n+1}(OCH_2CH_2)_xN(CH_3)_3]Cl$	12	0	0	1	1	0	0.00
			1	2.7	1.123	1.123	0.000	0.00
			2	4.9	1.162	1.162	0.000	0.00
6	$[C_nH_{2n+1}(OCH_2CH_2)_xNC_5H_5]Cl$	12	0	0	1	1	0	0.00
			1	1.65	1.097	1.098	0.001	0.09
			2	3.4	1.137	1.137	0.000	0.00
7	$[C_nH_{2n+1}(OCH_2CH_2)_xNC_5H_5]Cl$	16	0	0	1	1	0	0.00
			1	1.95	1.105	1.106	0.001	0.09
			2	4	1.148	1.148	0.000	0.00

On the basis of the obtained equations or figures (fig. 2) unknown *HI* values for other members of the considered series can be determined. Having known the actual number of EO groups in the considered surfactants, the equivalent number of oxyethylene groups in the standard series can be determined from figure 2 or from the equations derived above, and then the desired value of *HI* can be read from figure 1. The obtained values and literature date are shown in table 1.

The obtained results indicate that the proposed method can be successfully used to determine *HI* of surfactants only on the basis of *HI* values of two members of the considered homologous series. The absolute method error and the relative error are always less than 0.06 and 0.5 per cent, respectively. The mean absolute and relative errors are equal to 0.00075 and 0.07 per cent, respectively.

Summary

The new method of hydrophobicity index (*HI*) calculation is described. *HI* values of the standard series of surfactants and two members of the examined series are only needed. The mean absolute and relative errors of the method are equal to 0.00075 and 0.07 per cent, respectively.

Zusammenfassung

Es wurde eine neue Methode zur Bestimmung des Hydrophobieindexes (*HI*) angegeben. Zur Berechnung werden die *HI*-Werte der Standard-Tensidreihe und die Werte von zwei Gliedern der zu untersuchenden Tensidserie benötigt. Der mittlere absolute Fehler bzw. der mittlere relative Fehler dieser Methode betragen 0,00075 bzw. 0,07%.

References

1) *Lin, I. J., J. P. Friend, Y. Zimmels*, J. Colloid Interface Sci. **45**, 378 (1973).

2) *Lin, I. J., A. J. Metzer*, J. Phys. Chem. **75**, 3000 (1971).

3) *Jorne, J., E. Rubin*, J. Colloid Interface Sci. **38**, 639 (1972).

4) *Lin, I. J., L. Marszall*, J. Colloid Interface Sci. **57**, 85 (1976).

5) *Lin, I. J., B. M. Moudgil, P. Somasundaran*, Colloid & Polymer Sci. **252**, 407 (1974).

Author's address

Dr. *Jan Szymanowski*
Politechnika Poznańska
Institute of Chemical Technology
Poznań, Curie Sklodowskiej 1
Poland

Progr. Colloid & Polymer Sci. **63**, 99–104 (1978)

ISSN 0340-255 X

Department of Mineral Engineering, Technion – Israel Institute of Technology, Haifa (Israel) and Pharmacy No. 09068, Nowe (Poland)

Partition coefficient, HLB and effective chain length of surface-active agents

I. J. Lin and *L. Marszall*

With 3 figures and 4 tables

(Received November 3, 1976)

Introduction

In an earlier work, *Lin* and *Marszall* (1) described the relationship between CMC (critical micelle concentration) on the one hand, and hydrocarbon chain length and HLB (hydrophilic-lipophilic balance) on the other, and presented the results based on the concepts of effective chain length and hydrophobicity. The effects of structural modifications, and the correlation of HLB values derived from partition studies with the properties of the surfactant molecule, are examined in this paper.

HLB of surfactants

The HLB value represents the relative tendency of a surfactant molecule to transfer from an aqueous to a hydrocarbon environment and vice versa.

On the basis of the correlation found between the HLB and CMC values for hydrocarbon- and fluorocarbon surfactants, and of data on free energy change (ϕ) in transfer of the aliphatic chains from an aqueous to a non-polar phase, we obtain (2):

$$\mathrm{HLB} - 7 = \sum(\text{hydrophilic group numbers}) - n\phi'/(2.303\,kT) \quad [1]$$

where n is the number of $-CH_2-$ (or $-CF_2-$) groups in the straight-chain paraffinic molecule, $\phi' = \phi/n$, k – Boltzmann's constant, and T – absolute temperature. Hence, for a molecule with n carbon atoms, the group number per $-CH_2-$ or $-CF_2-$ group (see table 1) becomes:

$$\text{Hydrophobic group number per } -CH_2- = \phi'_{CH_2}/(2.303\,kT) = 0.475 \quad [2]$$

and

$$\text{Hydrophobic group number per } -CF_2- = \phi'_{CF_2}/(2.303\,kT) = 0.870 \quad [3]$$

For a given homologous series, the HLB decreases as the number of C atoms in the chain increases. For straight-chain compounds, n is the actual number of C atoms in the alkyl chain, while for branched compounds it is assumed to be n_{eff}; in terms of $-CH_2-$ or $-CF_2-$ groups, n_{eff} is that value of n which the analogous straight-chain compound must have to yield the same degree of surface activity as the branched one[1]).

Equation [1] therefore becomes:

$$\mathrm{HLB} - 7 = \sum(\text{hydrophilic group numbers}) - n_{eff}\phi'/(2.303\,kT). \quad [4]$$

Partition coefficient

The Nernst distribution law states that for partition of a surfactant in dilute solution between two immiscible solvents such as oil (o) and water (w)

$$K_p^m = C_o/C_w = \exp\,(-\Delta\mu^\circ/RT) \quad [5]$$

where the C's are equilibrium concentrations and K_p^m is known as the molar partition

[1]) Five methods for the determination of n_{eff} are described in (3).

Table 1. HLB group numbers for hydrophilic and lipophilic groups

Hydrophilic groups	Group number	Lipophilic groups	Group number	Derived groups	Group number
$-SO_4^- Na^+$	+38.7	$-CH_2-$	−0.475	$-(OCH_2CH_2)-$	+0.33
$-COO^-K^+$	+21.1	$-CH_3$	−0.475	$-(OCH_2CH_2CH_2)-$	−0.15
$-COO^- Na^+$	+19.1	$=CH-$	−0.475		
$-SO_3^- Na^+$	+11.0	$-CF_2-$	−0.870		
N(tertiary amine)	+ 9.4	$-CF_3$	−0.870		
Ester(sorbitan ring)	+ 6.8	Benzene ring	−1.662		
Ester (free)	+ 2.4				
−COOH	+ 2.1				
−OH (free)	+ 1.9				
−O− (ether group)	+ 1.3				
−OH (sorbitan ring)	+ 0.5				

coefficient. Eq. [5] may alternatively be expressed in terms of activities and

$$K_p^x = X_o/X_w = \gamma_w/\gamma_o \qquad [6]$$

where K_p^x is the partition coefficient given in mole fraction units, and related to K_p^m as

$$K_p^x/K_p^m = V_o/V_w \qquad [7]$$

V_o and V_w being the respective molar volumes of the phases.

The partition coefficient is important in processes such as zone refining, partition chromatography, and solvent extraction. It also plays a major role in surface chemistry and may be used to represent various physico-chemical equilibria.

The first extensive study of group contribution to K_p was undertaken by *Collander* (4, 5), who presented results for the partition of over 200 compounds (alcanoic acids, alkanols, alkyl amines) between water and diethyl ether, isobutanol, octanol and oleyl alcohol[2]). An interesting conclusion of that study (4) was that a logarithmic relationship exists between the partition coefficient of the same homologous series of compounds partitioned between water and two different solvent systems (O_1 and O_2):

$$\log K_{p(O_1)} = a \times \log K_{p(O_2)} + b \qquad [8]$$

where a and b are constants.

[2]) Other extensive lists of K_p values and a comprehensive review are given in (6, 7).

Another group-contribution study was undertaken by *Hansch* and *Fujita* (8, 9) who determined K_p of many hundreds of compounds for the 1-octanol-water system. The group contribution (π) was defined as

$$\pi_x = \log K_{p(RX)} - \log K_{p(RH)} \qquad [9]$$

where $K_{p(RH)}$ is the partition coefficient of the unsubstituted parent compound and $K_{p(RX)}$ that of the substituted derivatives. This equation is useful in studying the relationship between K_p and molecular structure.

The theoretical significance of the partition coefficient was discussed by *Flynn* (10), who claimed increasing evidence to the effect that when two structurally closely-related molecules are partitioned between the same two immiscible phases, the differences in free energy of transfer are directly related to the specific structural modification.

Although hydrocarbon chain length is known to be the major determining factor for the K_p of ionic and nonionic surfactants, we also have to take into consideration the effect of other structural parameters, e.g. the position of the polar head, double or triple bonds, aromatic groups, polar substitution and branching in the chain, cis- and trans-configuration, etc. In general, branching in an aliphatic chain reduces the K_p compared with the straight-chain isomer (7). A double bond can be assigned a negative π value, which again means a reduction in K_p. Some group-contribution values are listed in table 2.

The effect of $-CF_2-$ group is seen from eq. [9] and from the data listed in table 3. The

Table 2. Selected π values per group (7)

Group	π per group	Aliphatic group	π per group
$-CH_2-$	0.5	$-COOH$	-0.65
Branching in C chain	-0.2	$-OH$	-1.16
Double bond	-0.3	$-NH_2$	-1.16
Folding	-0.6	$-C{=}O$	-1.21
Intramolecular H bonding	$+0.65$	$-COOCH_3$	-0.27
		$-CN$	-0.84
		$-O-$	-0.98
		$-CONH_2$	-1.71
		$-F$	-0.17
		$-Cl$	$+0.39$
		$-Br$	$+0.60$

Table 3. Octanol/water partition coefficients (7) and transfer free energies of fluoro compounds

Compound	$\log K_p$	$\Delta \log K_p$ per CF_2	ϕ'_{CF_2}
CF_3CH_2OH	0.41		
$CF_3CF_2CH_2OH$	1.23	0.82	1.89 kT
$CF_3CF_2CF_2CH_2OH$	1.81	0.58	1.34 kT
$CF_3COOC_2H_5$	1.18		
$CF_3CF_2COOC_2H_5$	2.12	0.94	2.16 kT

$\Delta \log K_p$ per $CF_2 = \phi'_{CF_2}/(2.303\ kT)$

range obtained, 1.34 kT to 2.16 kT, is in fair agreement with literature data; *Phillips* (11) obtained 2.15 kT per $-CF_2-$ group for the adsorption energy calculated from CMC data, and *Lin* (2, 12) reported 2.0 kT. For ordinary surfactants the free energy change in transfer of a $-CH_2-$ group from an aqueous to an oil environment is estimated at 1.39 kT, as against 2.0 kT for a $-CF_2-$ group, i.e. $1.5\ \phi'_{CH_2} = \phi'_{CF_2}$. *Shinoda* et al. (13), in a study on micellization, similarly found that the CMC of a flourinated surfactant is close to that of an ordinary one whose hydrocarbon chain length is about 1.5 times that of the corresponding fluorocarbon.

Correlation between K_p and HLB

Schott (14) found a correlation between the HLB values of non-ionic surfactants and K_p. *Davies* (15), in a study on coalescence kinetics in emulsions, showed that provided the oil phase is nonpolar and nonreactive with the partitioning molecule, and disregarding the mutual solubilizing effect between the phases in the presence of the surfactant – the HLB is related to the partition coefficient as follows:

$$\mathrm{HLB} = 0.36 \times \ln (1/K_p) + 7 \qquad [10]$$

whence

$$\mathrm{HLB} - 7 = -0.36 \times \ln (K_p). \qquad [11]$$

Equation [11] allows calculation of HLB values from K_p values and vice versa if required.

Combination of eqs. [1] and [11] yields:

$$0.829 \times \log (K_p) = n\phi'/(2.303\ kT) - \sum (\text{hydrophilic group numbers}) \qquad [12]$$

which represents a linear plot with slope $\phi'/(0.829 \times 2.303\ kT)$, and an intercept $-\sum$(hydrophilic group numbers)/0.829.

The specific quantitative relationships between HLB, K_p and n for different homologous series of ionic surfactants are listed in table 4.

Figure 1 shows a typical curve of HLB vs. n, using data from table 1, and figure 2 – the partition coefficient of alkanoic acids plotted against alkyl chain length. The contribution of the methylene group is obtainable from the gradient of the regression line and that of the polar carboxyl group – from the intercept with the ordinate, which in turn depends on the solvent used. The free energy of transfer for the methylene group may be regarded as inde-

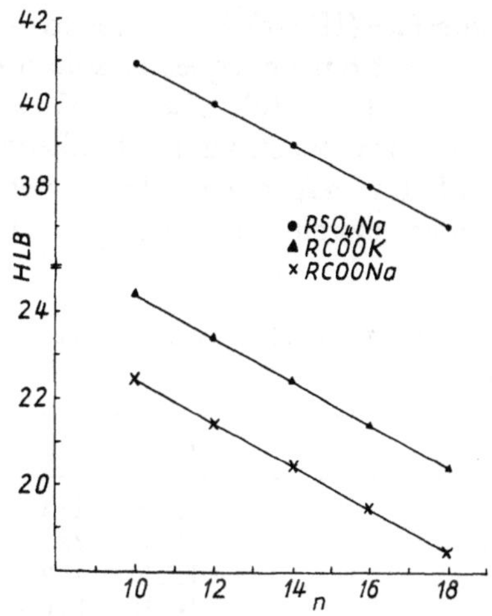

Fig. 1. Plot of HLB (eq. [1]) vs. chain length (n) for homologous series of ionic surfactants

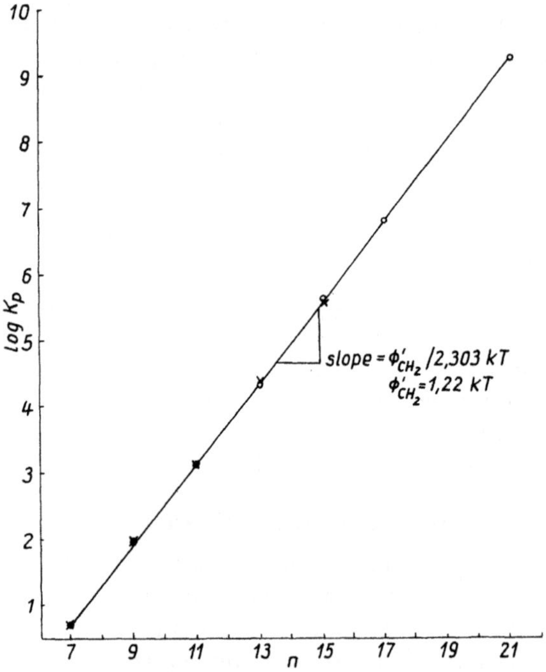

Fig. 2. Log K_p (in molar units) for alkanol acids vs. alkyl chain length (n). × after *Goodman* (23), ○ after *Smith* and *Tanford* (24).

pendent of the type of solvent, provided it is non-polar in nature (16) (see also the validity conditions for eq. [11]). In the case of polar solvents, the group contribution decreases with increasing polarity. In terms of free energy, the $-CH_2-$ group was found to have $-1.22\ kT$, as against the following literature data: (a) $\phi' = -1.39\ kT$ for complete transfer from an aqueous to a nonaqueous environment (17); (b) $\phi' = -1.39$ to $-1.41\ kT$ for transfer from an aqueous solution to an oil-water interface (18); (c) $\phi' = -1.34$ to $-1.7\ kT$ for transfer from an aqueous solution to a liquid state (19); (d) $\phi' = -1.33\ kT$ and $-1.37\ kT$ for partitioning of alkanols and alkyl sulfates respectively in a petroleum ether-water system. Free-energy data per $-CH_2-$ group and interfacial states are given in (19).

The relationships between K_p and n, listed in table 4 may be used for the calculation of the n_{eff} values of ionic surfactants with structural modifications (unsaturation, branching, folding, etc.). For example, one $-CH{=}CH-$ or $-C{\equiv}C-$ bond reduces n_{eff} by 0.52 $-CH_2-$ units, with a corresponding decrease in hydrophobicity. Similarly, branching reduces n_{eff} by 0.35 $-CH_2-$ units (or slightly more if the functional group is affected). Incidentally, a smaller HI ($=n_{eff}/n$) value is obtained for a shorter branch. Folding of an alkyl chain reduces n_{eff} by one $-CH_2-$ unit, and may be cause for nonlinearity of the K_p vs. n relationship (data presented in tables 2 and 4).

Crook et al. (20) studied the sub-CMC distribution of a series of p.t.-octylphenoxy-

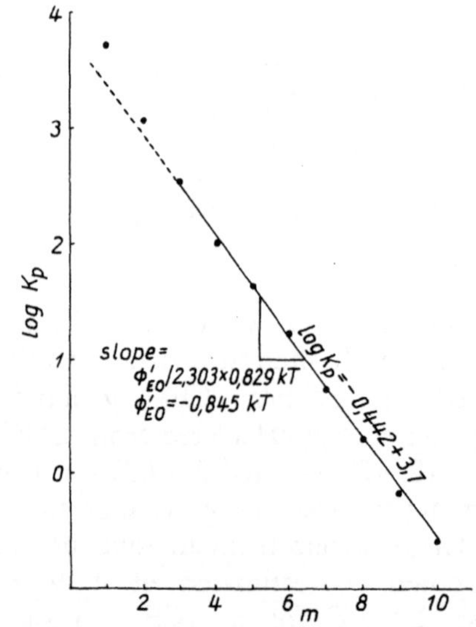

Fig. 3. Log K_p (in molar units) for p.t.-octylphenoxyethoxyethanols vs. ethylene-oxide chain length (m). After *Crook* et al. (20)

Table 4. The specific relationships between HLB, K_p and chain length for various homologous series of ionic surfactants

Surfactant	HLB vs. n Eq. [1]	K_p vs. n Eq. [12]
RCOOK	HLB = 28.1 − 0.475 n	log K_p = 0.573 n − 25.4
RCOONa	HLB = 26.1 − 0.475 n	log K_p = 0.573 n − 23.0
RSO_4Na	HLB = 45.7 − 0.475 n	log K_p = 0.573 n − 46.7
RSO_3Na	HLB = 18.0 − 0.475 n	log K_p = 0.573 n − 13.3
$C_nF_{2n+1}COOK$	HLB = 28.1 − 0.870 n	log K_p = 1.049 n − 25.4
$C_nF_{2n+1}COOH$	HLB = 9.1 − 0.870 n	log K_p = 1.049 n − 2.5

ethoxyethanols (OPE_{1-10}) in an iso-octane-water system at 25 °C. Figure 3 shows that K_p decreases as the ethylene-oxide (EO) chain length increases, with the attendant increase in hydrophilicity of the compound. The curve is linear for four orders of magnitude above $m = 3$; the non-linearity below that limit is due to interaction of the phenolic group and the short ethylene oxide side chains (1). On the other hand, *Petrov* and *Pozdnyshev* (21) found that addition of the first EO groups (depending on the size of the hydrocarbon radical) increases the hydrophobicity of an oxyethylated non-ionic surfactant (e.g. polyoxyethylated alkyl-phenol) and should favor a deviation toward higher K_p values.

The slope in figure 3 corresponds to a standard free energy change (ϕ_{EO}) of $-0.845\,kT$ (0.508 Kcal) per ethylene-oxide group. The group contribution for a functional group (e.g. a polar head at the end of an aliphatic chain) is obtainable by extrapolating the linear log K_p vs. n plot to zero n.

Finally, eq. [12] may be used for a pair of homologous series of compounds having different hydrophiles and the same lipophile; one contains propylene oxide ($-CH_2-CH_2-$ $-CH_2-O-$) and the other, ethylene oxide ($-CH_2-CH_2-O-$) groups, so that the intercept of log K_p vs. m is the same for both series. If the free energies ϕ' are additive, we have

$$\phi'_{PO} = 3\phi'_{-CH_2-} + \phi'_{-O-} \qquad [13]$$

and

$$\phi'_{EO} = 2\phi'_{-CH_2-} + \phi'_{-O-} \qquad [14]$$

then, subtracting eq. [14] from [13], we obtain:

$$\phi'_{PO} - \phi'_{EO} = \phi'_{-CH_2-} \qquad [15]$$

and eq. [15] yields ϕ' for the PO group:

$$\phi'_{PO} = 1.22\,kT + (-0.845\,kT) = 0.375\,kT. \qquad [16]$$

According to *Davies* (15) the group number per PO group equals -0.15, namely

$$-\phi'_{PO}/2.303\,kT = -0.375\,kT/2.303\,kT \cong -0.15\,. \qquad [17]$$

As indicated in recent studies (22), calculation of the HLB of polyoxypropylated surfactants on the basis of this group number and the nominal structure may lead to discrepancies. On the other hand, use of the same group number in conjunction with n_{eff} and HI yields correct HLB values for this type of surfactants and permits correlation of the CMC, K_p and HLB values.

Conclusions

The dual character of surfactants, reflected in their hydrophilic and hydrophobic moieties, has already been discussed with reference to solubilization, emulsifying power, micellization, wetting, hydration and adsorption. This paper deals with the effects of these properties on the partition coefficient.

The hypothesis that the thermodynamic properties of surfactants can be treated under a group-contribution approach has been well established, and methods are currently available for prediction of activity coefficients, partition coefficients, CMC, HLB, etc., from group values. Group methods have been employed extensively in determining structure-activity, and structure-HLB, relationships. By this means, a better insight is obtainable into solution and distribution tendencies of the surfactants and their functioning.

The n_{eff} contribution to the hydrophobicity index (HI) has been a notable advance and has allowed physical chemists to adopt a more rational approach to reagent design.

The ability to predict HLB values using K_p values is a considerable advantage. The HLB values have a fundamental significance in terms of free energy change per $-CH_2-$ or $-CF_2-$ group involved in chain transfer from the aqueous solution to the oil medium, should therefore be related to CMC, K_p, chain length and other thermodynamic properties.

Summary

The relation between the partition coefficient (K_p), hydrophilic-lipophilic balance (HLB) and effective chain length (n_{eff}) were investigated on the basis of literature data for ionic and nonionic surfactants. The effect of various types of structural modifications on the K_p and HLB of certain surfactants was considered. Using the n_{eff} concept for the substituted species, our approach is expected to make possible choice of surfactants for specific purposes in a less arbitrary fashion than up to now.

Zusammenfassung

Die Verhältnisse zwischen dem Verteilungskoeffizienten (K_p), dem hydrophil-lipophilen Gleichgewicht (HLB) und der wirksamen Kettenlänge (n_{eff}) wurden anhand von im Schrifttum vorhandenen Angaben über ionische und nichtionische Oberflächenwirkstoffe untersucht. Dabei wurden die Einwirkungen verschiedener Arten von strukturellen Modifikationen auf K_p und HLB bestimmter Oberflächenwirkstoffe betrachtet. Unter Zuhilfenahme des Begriffs der n_{eff} für die substituierten Gattungen hoffen die Verfasser, in der Lage zu sein, mit ihrem Vorgehen Oberflächenwirkstoffe für bestimmte Zwecke auf weniger willkürliche Art und Weise, als bisher möglich war, auswählen zu können.

References

1) *Lin, I. J., L. Marszall,* J. Colloid Interface Sci. **57**, 85 (1976).
2) *Lin, I. J.,* J. Phys. Chem. **76**, 2019 (1972).
3) *Lin, I. J., B. M. Moudgil, P. Somasundaran,* Colloid & Polymer Sci. **252**, 407 (1974).
4) *Collander, R.,* Acta Chem. Scand. **3**, 717 (1949); Ibid. **4**, 1085 (1950); Ibid. **5**, 774 (1951).
5) *Collander, R.,* Physiol. Plant. **7**, 420 (1954).
6) *Seidell, A.,* Solubility of Organic Compounds, 3rd Ed., Vol. II (New Jersey 1941).
7) *Leo, A., C. Hansch, D. Elkins,* Chem. Revs. **71**, 525 (1971).
8) *Hansch, C.,* Drug Design, Vol. I (New York 1971).
9) *Fujita, T., J. Iwasa, C. Hansch,* J. Amer. Chem. Soc. **86**, 5175 (1964).
10) *Flynn, G. L.,* J. Pharm. Sci. **60**, 345 (1971).
11) *Phillips, J. N.,* Trans. Faraday Soc. **51**, 561 (1955).
12) *Lin, I. J., J. P. Friend, Y. Zimmels,* J. Colloid Interface Sci. **45**, 378 (1973).
13) *Shinoda, K., M. Hato, T. Hayashi,* J. Phys. Chem. **76**, 909 (1972).
14) *Schott, H.,* J. Pharm. Sci. **60**, 648 (1971).
15) *Davies, J. T.,* Proc. 2nd Intern. Congr. Surface Activity, London **1**, 426 (1957).
16) *Davis, S. S., T. Higuchi, J. H. Rytting,* J. Pharm. Pharmac. **24**, 30P (1972).
17) *Mukerjee, P.,* Advan. Colloid Interface Sci. **1**, 241 (1967).
18) *Haydon, D. A., F. H. Taylor,* Phil. Trans. Roy. Soc., London **252**, 225 (1960).
19) *Lin, I. J., P. Somasundaran,* J. Colloid Interface Sci. **37**, 731 (1971).
20) *Crook, E. H., D. B. Fordyce, G. F. Trebbi,* J. Colloid Sci. **20**, 191 (1965); Ibid., J. Phys. Chem. **67**, 1987 (1963).
21) *Petrov, A. A., G. N. Pozdnyshev,* Kolloidn. Zh. **28**, 858 (1966).
22) *Lin, I. J., L. Marszall,* Tenside Detergents, **14**, 131 (1977).
23) *Goodman, D. S.,* J. Amer. Chem. Soc. **80**, 3887 (1958).
24) *Smith, R., C. Tanford,* Proc. Nat. Acad. Sci., USA **70**, 289 (1973).

Authors' addresses:

Dr. *L. Marszall*
Pharmacy No. 09068, Rynek 12,
PL–86–170 Nowe k/Swiecia (Poland)

Dr. *I. J. Lin*
Department of Mineral Engineering,
Technion – Israel Institute of Technology,
Haifa (Israel)

Progr. Colloid & Polymer Sci. **63**, 105–111 (1978)

ISSN 0340-255 X

Department of Chemistry, University College of Science, Calcutta (India)

Correction to microscopically determined particle size according to diffraction theory*)

Application to oil-in-water emulsion droplets and aqueous AgI sol particles

M. Sengupta, S. S. Jana, and *D. N. Biswas*

With 7 figures

(Received January 1, 1977)

Introduction

In microscopic scanning of isolated objects of size less than 1 or 2 microns for the purpose of their size determination, the effect of diffraction of light due to the finite aperture of the imaging system must be taken into account, even though the appearance of the boundary in the image is apparently distinct. For this, the theoretical equations giving the light intensity distribution in the image, for particles of varying degrees of contrast with respect to the surround, as also for different kinds of illumination of the particles, have to be used.

In our recent experiments designed to measure the electrophoretic mobilities of varying-sized particles of AgI as also of different oil-in-water emulsions, the size of the particles was measured microscopically using the combination: 20× objective and 28× eye piece, fitted with an ocular micrometer, the illumination (visible light) being controlled by a sub-stage condenser. The observed particle sizes were corrected by applying the diffraction correction mentioned above, after taking into account properly the varying degrees of contrast of the different types of particles with respect to the surround, as also the degree of coherence of the illumination actually used, as determined by the numerical apertures of the condenser and the objective, and their relative placement in the microscope system. Since the method of microscopic measurement of particle size is used quite often in colloid-chemical work, so the method of size correction mentioned may prove to be interesting and useful in other contexts. It is the purpose of the present communication to describe the method of size correction used.

Diffraction theory for opaque particles

The diffraction theory for calculating the light intensity distribution in the image of disc-shaped particles, located in an uniform surround and axially illuminated (Köhler) with unpolarised monochromatic light, has been worked out by *Osterberg* and *Pride* (1) for the case when the numerical aperture of the condenser was taken as almost equal to zero (fully coherent illumination); and by *Osterberg* and *Smith* (2) for the case when the numerical apertures of the condenser and the objective are same (one special case of partially coherent illumination). In the second case mentioned above, which is of interest to us, the solution contains two infinite sums I_{ud} and I_{dd} which depend upon the radius of the particle (u), and the distance from the centre of the diffraction image (r) in the image plane, both in reduced units. These functions have been computed and tabulated (3) for values of u and r in Airy units (u' and r'), over the ranges: $u' = 0.2 - 2.0$ and $r' = 0 - 2.0$.

*) Presented at the Golden Jubilee Celebration Symposium of the Department of Applied Physics, Calcutta University (Jan. 1976).

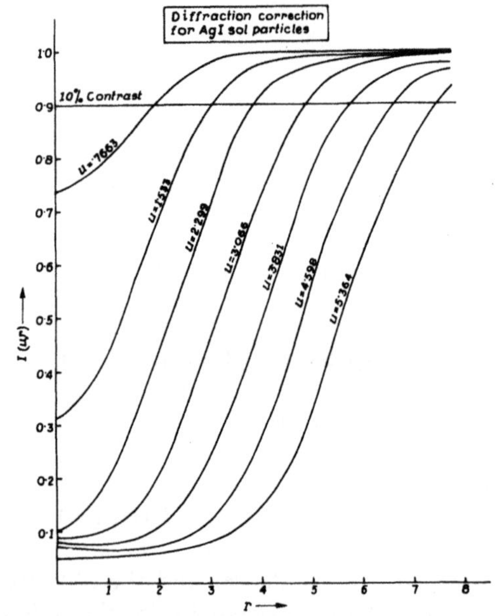

Figs. 1–5. Intensity distribution vs. reduced radial distance, in the image, for some different reduced radius values, for spherical particles in aqueous suspension. 1: AgI particles; 2: Iodobenzene droplets; 3: Castor oil droplets; 4: Nujol droplets and 5: Dimethylaniline droplets

In the case of opaque particles, when the ratio of the amplitude transmittance of the particle to that of the surround is equal to zero, the equation of *Osterberg* and *Smith* for the light intensity in the image I, reduces to:

$$I(u, r) = 1 - \frac{2I_{ud}}{\pi} + \frac{I_{dd}}{2\pi^2}. \quad [1]$$

The graphical plots of $I\,(u, r)$ vs. r for eight different values of u' (0.2, 0.4, 0.6, 0.8, 1.0, 1.2, 1.4 and 2.0) have been given by *Smith* (3); we reproduce it here for the sake of easy reference (fig. 1)[1]).

In actual practice the particles observed microscopically are usually spherical in shape, and not disc-like. Unfortunately, the theoretical treatment for diffraction in the case of spherical particles is not yet available in a final form. We therefore have to make the assumption that the effect of diffraction for spherical particles is similar to that of disc-shaped particles, and then utilise the theoretical results available for the latter case.

Diffraction theory for transparent particles

The diffraction theory for light intensity distribution in the image of disc-shaped particles of varying contrast with respect to an uniform surround (i.e. of arbitrary phase and amplitude transmission), partially coherently illuminated axially, has been worked out by *De* and *Som* (4, 5), for four distinct cases of degree of coherence of the source, namely: $\varrho \to 0$, $\varrho < 1$, $\varrho \geqslant 1$ and $\varrho \to \infty$, ϱ being the ratio of the numerical apertures of the condenser to that of the objective. Putting the amplitude transmissions of the medium and the particle equal, i.e. for the case of pure phase object, the expression for the light intensity distribution for the case of interest to us, namely $\varrho \geqslant 1$, has the form (4):

$$I(r) = 1 + 32\,\mathrm{Sin}^2\left(\frac{\Delta}{2}\right)\left(A - \frac{B}{2}\right) \quad [2]$$

where A and B are complicated functions of u and r^2); Δ is related to the (average) optical path difference, δ, by the relation: $\Delta = \frac{2\pi}{\lambda}\,\delta$, δ itself being given by: $\delta = \frac{\pi u}{2}(n - n_0)$, n and n_0 being the refractive indices of the materials constituting the dispersed and dispersion medium respectively; u the actual geometrical radius of the particles; and λ the wavelength of the light used.

Numerical values of the functions A and B, for various values of 'reduced' radial distance in the image (r) and the reduced actual radius of the particle (u) can be obtained from the

---

1) The plots have been freshly made using tabulated values of the functions I_{ud} and I_{dd}, and designated for different values of u and r, rather than of u' and r'.

2) The functions A, B correspond respectively to the double sum and single sum functions:

$$\sum_{p=1}^{\infty}\sum_{q=1}^{\infty} p \cdot q \phi_{p,q}(\sigma, \omega') J(p, q, \sigma\varrho)$$

and

$$\sum_{p=1}^{\infty}\sum_{q=1}^{\infty} p \cdot \phi_{p,q}(\sigma, \omega')$$

of *Som*'s paper (eqn. 18b, ref. 4).

values computed and tabulated by *Som* (5)[3]. Using these values, and the value of δ corresponding to the particular range of (i) u values and (ii) the n value of interest, and (iii) the λ value of the light used (mean value for white light), graphical plots of I as function of r for some different values of u can be constructed (figs. 2–5).

Application to the cases of (i) solid AgI particles and (ii) liquid oil droplets, such as nujol, castor oil, iodobenzene and dimethylaniline, all in aqueous suspension

In our microscope assembly a beam of white light reflected from a concave mirror enters into the condenser, which was a simple convex lens of diameter 2.5 cm. The distance from this lens surface to the object plane being 5.2 cm, the numerical aperture of the condenser was $\sim$0.4. Also, the numerical aperture of the objective was 0.4, so that in our experiments the degree of coherence of the illumination used was $\sim$1.

(i) Opaque particles (AgI): The AgI particles can very reasonably be assumed to be perfectly opaque, and hence this case can be treated by using the plots for the corresponding case (fig. 1). Assuming that in usual visual observations, the boundary of the image is discernible from the background upto a reduction to about 10% of the peak light intensity in the image ("10% contrast") (6), the such values of r (r_0, say) corresponding to different values of u (u_0, say) can be read off from figure 1, and used to construct the plot (fig. 6), giving for any value of the reduced actual radius of the particles (u_0), the reduced distance (r_0) corresponding to 10% contrast in the image, i.e., the reduced apparent size of the image.

Further, the particle radius is related to the corresponding 'reduced' quantity by a simple equation involving the wavelength of the light used (λ), and the numerical aperture of the objective (N.A.). For the case when the diffraction correction has yet to be made, e.g. for the observed particle radius values in microns, this equation gives:

$$\text{(Reduced) obsd. radius} = \text{obsd. radius} \frac{2\pi}{\lambda} \text{(N.A.)} \qquad [3]$$

where $\lambda_{av.}$ for the light used = 0.55 microns, and N.A. for the objective used = 0.40; and again, for the case when the necessary correction has been made, e.g. after using figure 6, the same equation gives:

$$\text{Corrected radius} = \frac{\text{(Reduced) corrected radius}}{\frac{2\pi}{\lambda}\text{(N.A.)}} \qquad [4]$$

(ii) Transparent particles: In the case of the droplets of the different oils the refractive index values were obtained from literature (7): water: $n = 1.333$, nujol: $n = 1.44$, iodobenzene: $n = 1.621$, castor oil: $n = 1.471$ and dimethylaniline: $n = 1.558$[4]). Further, $\lambda_{av.}$ being 0.55 microns, and the range of particle size of interest being 1–6 microns, graphical plots of I vs. r for 4–6 different u values were constructed (figs. 2–5).

For small particles (u small), these intensity distribution curves are similar to those obtained in the case of opaque particles, showing that for small particles, the nature of the correction becomes essentially identical in the two cases. The further method of size correction using these curves, involving '10% contrast estimation' (fig. 6, curves for the different oils),

[3]) The values of the double and single sum functions for selected values of u, ϱ and r are computed and tabulated in ref. (5). The values of B_1, B_2 of tables 1, 2, 3 (for the case $\varrho = 1$) give the values of our A function; likewise, the values of B_1, B_2 (for the case $\varrho = \infty$) give the values of our B function. Further, as *Som* (5) has made computations for annular objects with outer diameter equal to three times the inner diameter, each set of his computed results (tables 1, 2, 3 of ref. (5)) makes possible the evaluation of the functions A and B for two different values of particle diameter (one, three times the other). Thus table 1 gave A, B for $u = 1$ and 3, table 2 gave for $u = 1.5$ and 4.5 and table 3 gave for $u = 2$ and 6.

[4]) The experimental (ambient) temperature, being not very different from the temperature (25 °C) for which the literature values are recorded, any small correction for this small temperature difference involved was neglected.

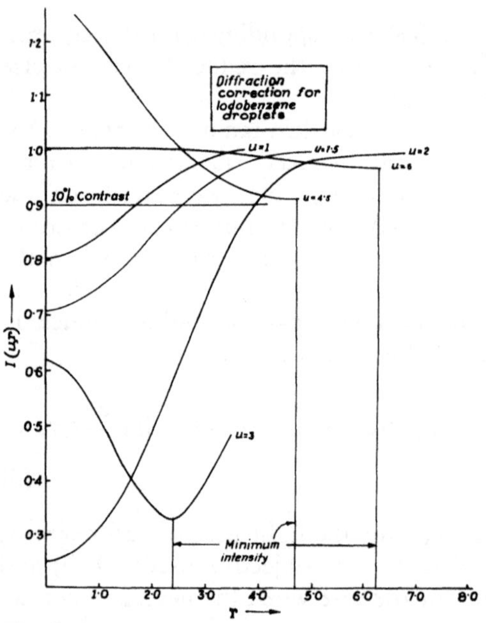

Fig. 2. (vide p. 106)

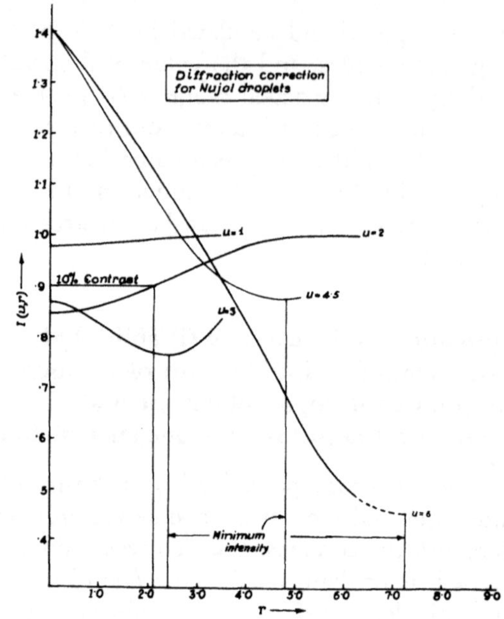

Fig. 4. (vide p. 106)

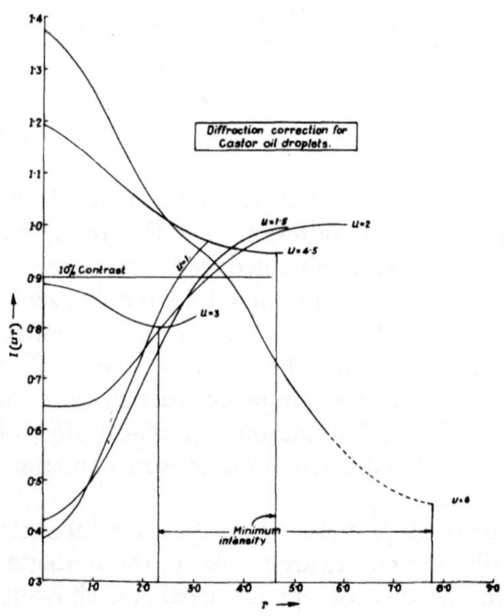

Fig. 3. (vide p. 106)

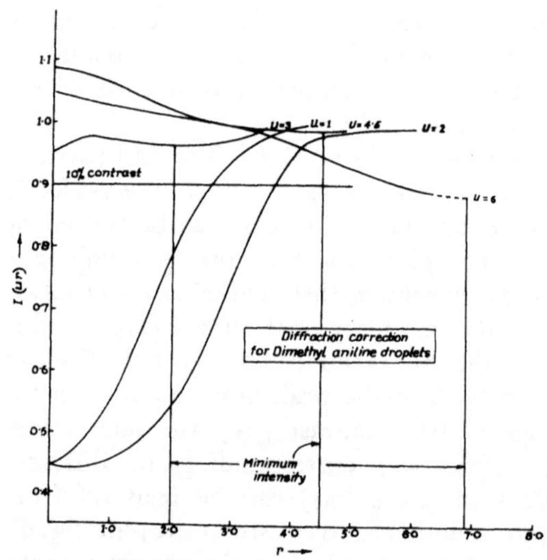

Fig. 5. (vide p. 106)

is similar to that described earlier, and need not be further described[5]).

[5]) In the case of nujol particles, only the intensity distribution curve for $u = 2$ lends itself to 10% contrast estimation process. Neither the curve for $u = 1$ which gives an almost uniform bright radial intensity distribution in the image, nor even the curve for $u = 1.5$ [which surprisingly falls below those for $u = 1$ and 2, (and would lead to an opposite "r_0 (10% contrast) vs. u_0" dependence if it were used for drawing that plot in conjunction with the $u = 2$ curve), and hence was neither considered further, nor is shown here] is thus amenable. Hence the plot "average size at 10% contrast vs. reduced radius" for nujol cannot be constructed. Fortunately, however, in this system the only solitary datum for small particle corresponded to an u value sufficiently close to $u = 2$, so that the (u vs. r) curve corresponding to $u = 2$ could directly be used for "10% contrast" size estimation. Also, the datum for the next larger u value, as also all the rest, could confidently be referred to the r_{min} vs. u_0 curve drawn from r vs. u curves for $u = 3$, 4, 5 and 6. In the case of dimethyl-aniline emulsion also, the curve for $u = 1.5$ lies above those for $u = 1$ and 2, and does not lend itself to be used for constructing the r_0 (10% contrast) vs. u_0 curve. This curve is also not shown here.

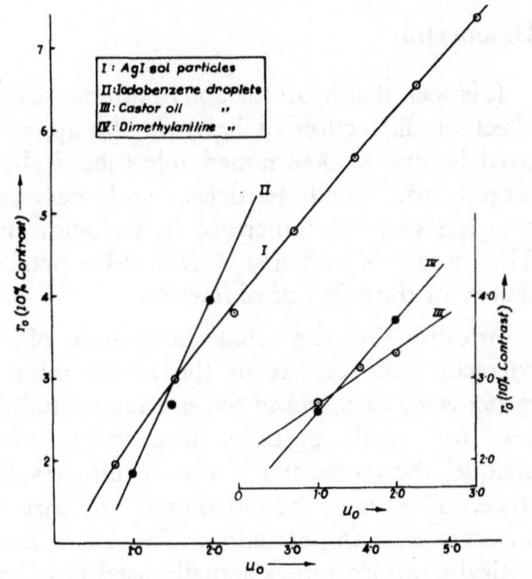

Fig. 6. Reduced apparent size at 10% contrast vs. reduced radius of the particle

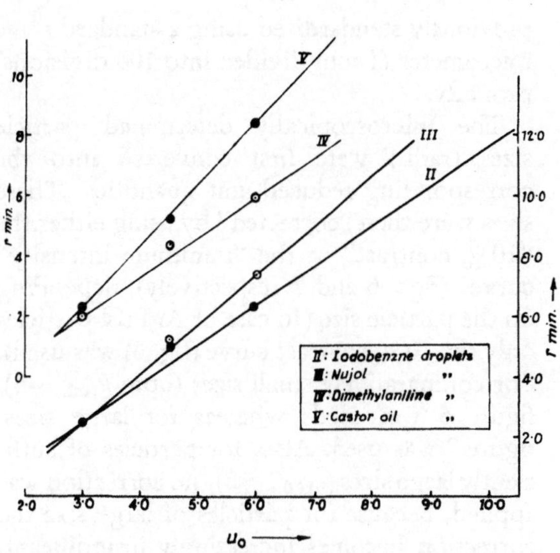

Fig. 7. Reduced apparent size at minimum intensity vs. reduced radius of the particle

For comparatively larger particles ($u(\geqslant 3)$ large), on the other hand, the intensity distribution curve for transparent particles shows a minimum[6]), implying that the image contains a central bright zone and a less bright periphery separated by a darker ring. Images of this nature were actually observed in the case of comparatively larger particles in iodobenzene and nujol emulsions. In such cases the dark periphery was naturally taken to signify the boundary of the particle, a process which would imply that in this case, the position of the minimum in the intensity distribution curve (r_{min}, say) be considered to have given the position of the particle boundary (u_0). Hence u_0 was here plotted against r_{min} (for u large, for all the different systems 'namely' nujol, castor oil, iodobenzene and dimethylaniline systems) to give the curves shown in figure 7.

[6]) The thus constructed intensity vs. reduced radial distance curves corresponding to high values of u (particularly, $u = 6$) in case of almost all the four systems considered, show that the curves, similarly as those for smaller values of u, do approach towards a minimum. However, the exact position of the minimum has to be surmised somewhat (see figs. 2–4), in absence of sufficient numerical results in literature for completing these curves.

The rest of the procedure was similar to that already described. After converting the (observed) radii into the reduced (observed) radii by the equation [3], the reduced (corrected) radii were estimated with the help of these curves (figs. 6, 7), and finally from these the (corrected) radii were computed by using equation [4].

Results of size correction in some actual experimental cases

The AgI sol and the different oil-in-water emulsions were prepared by the method described elsewhere (8, 9, 10). The sol particles were apparently almost spherical in shape, though some irregular agglomerations were also present. The emulsion droplets were all spherical. Only the isolated individual particles of varying sizes were singled out for microelectrophoretic mobility measurements; the sizes werde determined by temporarily holding the particles stationary in the field of the "fine" ocular micrometer (1 smal div. effectively equal to 5.25×10^{-4} cm in the object plane). The eye-piece had another graticule ('coarse') mounted in it, this was used for timing the electrophoretic movement of the particles. Both the eye-piece micrometers were previously standardised using a standard stage micrometer (1 mm divided into 100 divisions) properly.

The microscopically determined particle sizes (radii) were first converted into the corresponding reduced-unit quantities. These sizes were then "corrected" by using either the "10% contrast" or the "minimum intensity" curves (figs. 6 and 7 respectively), depending on the particle size (In case of AgI sol particles only the 10% contrast curve (fig. 6) was used). For comparatively small sizes (upto $u_{red.} \sim 4$), figure 6 was used; whereas for large sizes, figure 7 was used. Also, for particles of sufficiently large sizes ($u_{red.} > 8$), no correction was applied, because for particles of large size the correction becomes increasingly insignificant.

In the following table are shown some typical corrected and uncorrected particle sizes, over the generally scanned particle size range, for (*a*) droplets of two different oil emulsions in distilled water: (i) iodobenzene ($n = 1.621$) and (ii) nujol ($n = 1.44$); and (*b*) particles of AgI sol. Whereas the uncorrected radii in the first two cases are those actually measured visually, in case of the AgI sol the settling tendency of the particles increasing very rapidly with even slight increase in particle size, only about one micron (radius) size particles were studied for microelectrophoresis and simultaneous size measurements. The particle size range shown scanned in the table for the case, gives merely the anticipated size correction as would be inferred from figure 6 for particles over this size range.

Discussion

It is seen that in all the above three cases the effect of diffraction of light on the apparent particle size as determined microscopically is largest for small particles, and decreases progressively with increase in particles size. This result is anticipated from the detailed theory of the effect of diffraction.

Additionally, the actual magnitude of the correction as seen from the above table is rather large throughout the size range studied, and for small particles (especially, when opaque) the correction is almost unbelievably large. This may be attributed in part to observational imperfection. The 20 × 28 × optical combination as actually used is at best a poor combination to use for verification of refined diffraction correction theories. The resolution accuracy is consequently much too low (0.2 scale division of the eyepiece micrometer, equivalent to 0.5 micron in the object plane). A more effective optical arrangement with an overall 100 × objective magnification, in conjunction with the use of a still finer ocular micrometer (1 div. = 0.5 micron, with corresponding increase in resolution accuracy) is being developed. Further, the 10% contrast criterion as used for drawing the theoretical r_0 (10% contrast) vs. u_0 curves is only a convenient *ad hoc* device for delineation of the image boundary; its rigorous observance in actual visual experiments can hardly be expected to be possible in a perfectly objective manner. Also, the use of the average wave

Size correction for oil-in-water emulsion droplets and AgI sol particles

	Iodobenzene ($n = 1.621$)				Nujol ($n = 1.44$)				AgI			
	Uncorr. radii		Corr. Radii		Uncorr. radii		Corr. radii		Uncorr. radii		Corr. radii	
Micrometer scale divs.	Micron	Red. Unit	Red. Unit	Micron	Micron	Red. Unit	Red. Unit	Micron	Micron	Red. Unit	Red. Unit	Micron
0.1	0.525	2.400	1.30	0.284	0.525	2.400	2.29	0.500	0.4	1.828	0.70	0.153
0.15	0.787	3.599	1.85	0.405	0.787	3.599	3.43	0.750	0.6	2.741	1.30	0.285
0.20	1.050	4.798	2.42	0.530	1.050	4.798	4.80	1.000	0.8	3.656	2.07	0.453
0.25	1.313	5.995	5.75	1.258	1.313	6.001	5.22	1.142	1.0	4.570	2.87	0.627
0.35	1.838	8.401	7.65	1.674	1.838	8.401	6.75	1.477	1.4	6.398	4.47	0.978
0.40	2.130	9.711	8.65	1.892	2.130	9.734	7.52	1.645	1.6	7.312	5.30	1.159
0.45	2.363	10.80	9.50	2.079	2.362	10.80	8.20	1.795				

length value for the white light used for illumination, in absence of a monochromatic source, is again an approximation.

The approximations involved in the theoretical treatment of diffraction correction must also be mentioned in this connection. Whereas the theory is developed for the case of disc-shaped particles, it has been applied to spherical particles. (In the case of solids like AgI, the geometry may vary from particle to particle, and deviate considerably from spherical; hence also, likely larger anomalies in the result).

Whether some or all of the above mentioned effects together are able to account for the rather large particle size corrections as found in this preliminary study can only be decided on the basis of more refined experimental results. Further work in this line is in progress.

Acknowledgement

Our sincere thanks are due to Dr. *S. C. Som* of the Department of Applied Physics (Optics Section), Calcutta University, for helpful discussions throughout the course of this work.

Summary

The correction to be applied to the apparent particle size as determined microscopically, on account of diffraction of light, has been considered. Two distinct cases viz. amplitude (opaque) and phase (transparent) particles have been considered, the theoretical treatment for the diffraction of light for which cases was available from literature. The treatment has been applied for particle size correction in case of spherical water-suspended particles of silver iodide, nujol, castor oil, iodobenzene and dimethylaniline.

Zusammenfassung

Die Arbeit behandelt die Korrektur, die aufgrund der Lichtbrechung bei der mikroskopischen Bestimmung der scheinbaren Teilchengröße anzuwenden ist. Das Verfahren wurde auf Teilchengrößenkorrekturen für kugelige, in Wasser suspendierte Teilchen von Silberjodid, Nujol, Castoröl, Jodbenzol und Dimethylanilin angewandt.

References

1) *Osterberg, H., G. E. Pride*, J. Opt. Soc. Am. **40**, 14 (1950).

2) *Osterberg, H., L. W. Smith*, J. Opt. Soc. Am. **50**, 362 (1960).

3) *Smith, L. W.*, J. Opt. Soc. Am. **50**, 369 (1960).

4) *De, M., S. C. Som.*, J. Opt. Soc. Am. **53**, 779 (1963).

5) *Som, S. C.*, J. Opt. Soc. Am. **57**, 1499 (1967).

6) *Moon*, The Scientific Basis of Illuminating Engineering, p. 420, Dover Pub. (New York 1961).

7) International Critical Tables, Vol. I (National Acad. Sc.) (New York 1926).

8) *Jana, S. S., M. Sengupta, D. N. Biswas*, Progr. Colloid & Polymer Sci. **63**, 78–86 (1978).

9) *Jana, S. S.*, Ph. D. (Science) Thesis, Calcutta University (1973).

10) *Biswas, D. N.*, Ph. D. (Science) Thesis, Calcutta University (1975).

Authors' address:

M. Sengupta
Department of Chemistry,
University College of Science
Calcutta-700009
(India)

Progr. Colloid & Polymer Sci. **63**, 112–115 (1978)

ISSN 0340-255 X

Department of Chemistry, Aligarh Muslim University, Aligarh (India)

Micellar behaviour of sodium dodecyl sulfate in presence of a cationic dye

H. N. Singh, Sahab Singh, and *Onkar Singh*

With 5 figures and 1 table

(Received July 7, 1976)

Introduction

It has been known for more than three-fourth of a century that indicator dyes have their colours altered when present in solutions along with soaps or detergents. The interaction of dye with surfactants leading to colour change has been attributed to novel effect which makes the detergents differ from other electrolytes (1) to the formation of ionpairs (2) of micelles (3) of an insoluble complex (4, 5) the presence of dyedimers and high aggregates with one or more soap molecules (4) of insoluble dye-detergent complex and induced micelles (6). The formation of a very insoluble salt between ionic dyes and oppositely charged detergent is a very general phenomenon (9). Many dyes in the above reference have been taken as a source of determination of cmc but, other shown a strong interaction with surfactant micelles.

It is a well known fact that dyes possessing a charge opposite to that on the surfactant micelles show strong interaction to form dye-detergent complex, but no such interactions are reported with the dyes and detergent micelles having similar charges (7, 8, 9, 10).

The object of the present investigation has been to examine the interaction of a triphenyl methane (fuchsin basic) dye as a tool to determine the cmc of surfactants in water. This dye was found to lose intensity in presence of very dilute solution of SDS which further increased at higher concentration.

Materials

The SDS was purchased from Sigma Chemical Company, U.S.A. It was recrystallized thrice from ethanol and kept over petroleum ether for 100 hours to remove traces of free alcohol. The substance was dried in vacuum before use.

The cetyl trimethyl ammonium bromide was obtained from British Drug House (England). It was purified by extraction with ether and recrystallization from methanol, repeated thrice.

Method

Standard solutions of dye and detergents were prepared by weighing. In experiments where no additional electrolyte was used, a stock solution of the detergent above the cmc was made in a dye solution and this was added to the same dye solution in water or phosphate buffer of known pH to get various concentrations of the detergent. Thus the dye concentration was constant for a given set of experiments. Where sodium chloride was used the stock solutions of dye and of detergent (above cmc) were prepared in a desired concentration of the salt. Solutions of various concentrations of detergent with fixed concentration of dye and NaCl were prepared by suitable dilutions.

For conductivity measurements, distilled water equilibrated with laboratory atmosphere was used.

A Bausch and Lomb spectronic-20 and 1 cm cuvettes were used for optical determinations. Conductivity measurements were carried out with a Philips conductivity bridge model PR 9500 using platinized conductivity cell type PV 9055. The unit incorporates a Wheatstone bridge with measuring frequencies 50 c/s and 1000 c/s.

Results and discussion

The variation of optical density with dye concentration at 535 nm is shown in figure 1. The deviations from Beer's Law at very low concentrations indicate the association of dye in water (6). Hence in the present investigation, the concentration of the dye was always taken 2×10^{-6} M or higher.

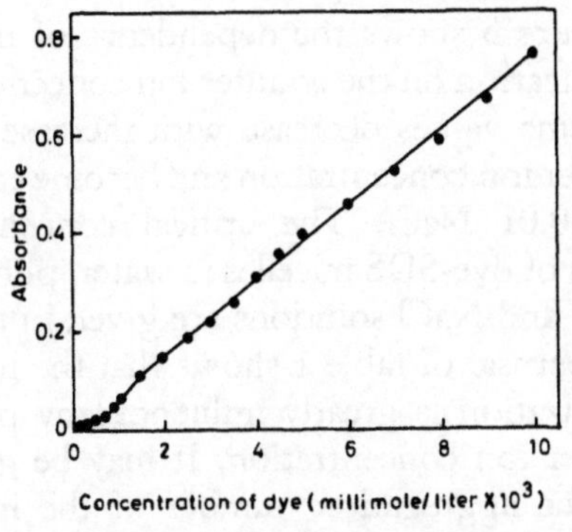

Fig. 1. Variation of absorbance with dye concentration

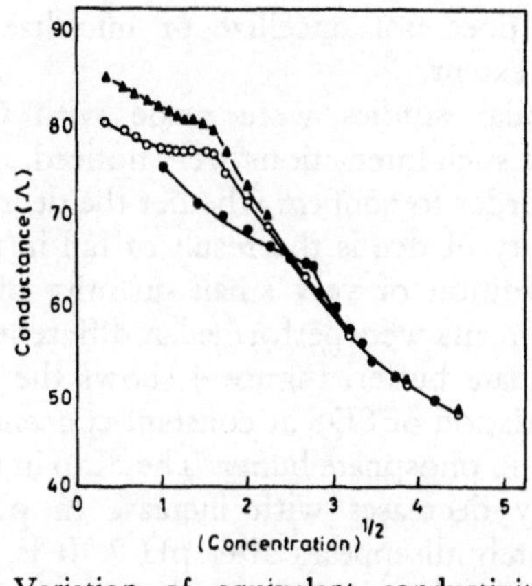

Fig. 3. Variation of equivalent conductivity with square root of SDS concentration curve. -●- without dye, -○- with 0.002 mMl^{-1} dye, -△- with 0.004 mMl^{-1} dye

Spectra were recorded at constant concentration of dye in presence of SDS below and above cmc. No colour change or spectral shift was noticed in any case. Figure 2 shows the deviation from Beer's Law with increase in the concentration of SDS at constant concentration of dye in water. The optical density first falls and then gradually increases with increase in the surfactant concentration and reaches a constant value after 2.5 millimoles of SDS. The concentration of SDS after which there is no change in optical density is taken as the cmc. The gradual increase in optical density and the lower value of SDS in water clearly indicates the incorporation of dye with the surfactant micelles.

The correctness of this interpretation may be provided by comparing colorimetric and conductometric measurements on the same system containing both dye and detergent. Figure 3 shows the results for SDS in the presence of 2×10^{-6} M and 4×10^{-6} M of fuchsin. It may be seen that equivalent conductivity shows first deviation from linearity near a concentration when the optical density begins to increase (fig. 3). The equivalent conductivity then decreases showing the formation of micelles, with a definite kink at the point where the optical density becomes constant (fig. 2).

It is clearly apparent that the drop in equivalent conductivity and hence micellization occurs at markedly lower concentrations in the presence of dye. The above facts show the existence of micelles induced by the presence of dye at concentrations at which the detergent

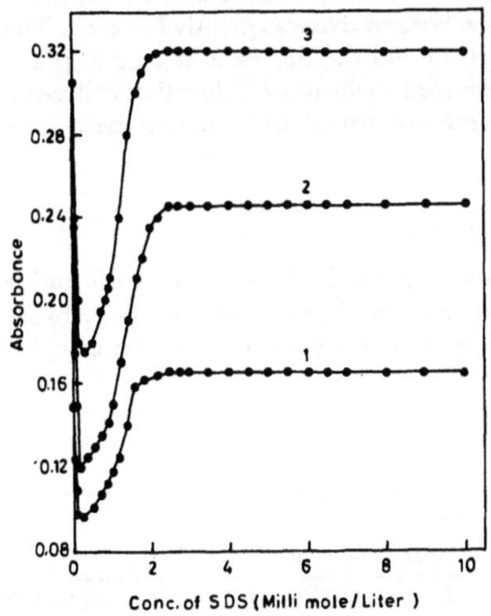

Fig. 2. Variation of absorbance with SDS concentration. 1) 0.002 mMl^{-1} dye, 2) 0.003 mMl^{-1} dye, 3) 0.004 mMl^{-1} dye

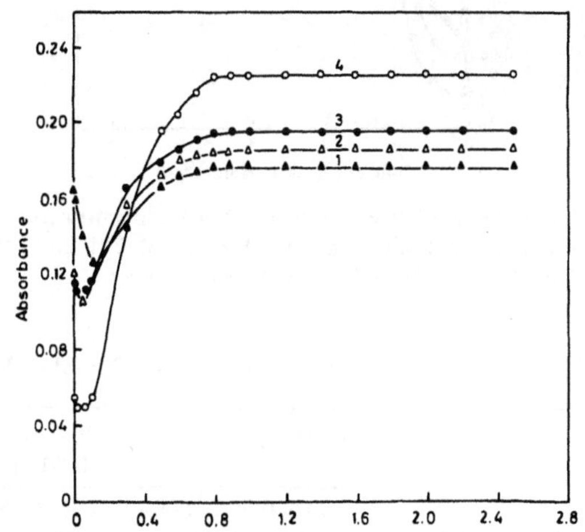

Fig. 4. Dependence of cmc on pH. 1) pH 6.2, 2) pH 6.95, 3) pH 7.13, 4) pH 7.93

alone does not micellize or micellizes to a lesser extent.

Similar studies were made with CTAB, but no such interactions were noticed.

In order to confirm whether the decrease in intensity of dye is the result of fall in pH on the addition of very small quantity of SDS, experiments were performed at different pH in phosphate buffer. Figure 4 shows the results on variation of SDS at constant concentration of dye in phosphate buffer. The drop in optical density decreases with increase in pH and ultimately disappears after pH 7. It is, therefore, clear that the decrease in intensity of the dye is not the result of change of pH due to the addition of SDS, but it may be solely the effect of pH on the ionic character of the dye. Hence it is apparent from figure 4 (Curve 4) that a gradual increase in optical density is observed in buffer at higher pH than 7. This clearly indicates that the dye molecules begin to associate at very low concentrations of SDS to form induced micelles. The induced cmc values in phosphate buffer, though sufficiently lower than in water, are independent of pH.

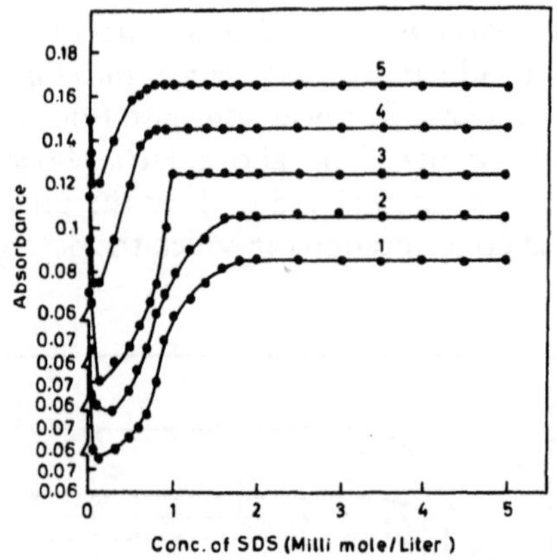

Fig. 5. Dependence of cmc on NaCl concentration in constant 0.002 mMl^{-1} of dye. 1) 0.001 M, 2) 0.0025 M, 3) 0.005 M, 4) 0.01 M, 5) 0.1 M of NaCl

Figure 5 shows the dependence of induced micellization on the counter ion concentration. The cmc values decrease with increase in the counter ion concentration and become constant after 0.01 NaCl. The critical concentration values of dye-SDS micelles in water, phosphate buffer and NaCl solutions are given in table 1.

A perusal of table 1 shows that the induced micellization is greatly influenced by pH and counter ion concentration. It may be guessed that the aggregation number of the induced micelles acquires a constant value at 0.01 M NaCl and over entire pH values in phosphate buffer studied. The present dye could not be used as a suitable tool for the determination of cmc of surfactants but it reflects important information regarding the interaction of anionic surfactants leading to induced micellization.

Further investigations on the determination of aggregation number, viscometric and tensiometric measurements are in progress.

Acknowledgement

Council of Scientific and Industrial Research, New Delhi, is gratefully acknowledged for financial support. Thanks are also due to Professor *W. Rahman* for research facilities.

Summary

Dye adsorption spectroscopy has been used as a tool to determine cmc of anionic and cationic surfactants in water. The effect of pH and sodium chloride was examined on cmc. The dye was found to interact with sodium dodecyl sulfate (SDS) micelles but did not interact with cetyltrimethyl ammonium bromide (CTAB). In phosphate buffer and NaCl, the cmc of SDS in presence of dye was greatly lowered. The mode of interaction and the decrease in cmc in presence of dye is discussed in terms of induced micellization. The results were confirmed by conductometric measurement.

Zusammenfassung

Zur Bestimmung der CMC anionischer und kationischer Tenside in Wasser wurde die Absorptionsspektroskopie bei Gegenwart eines Triphenylmethan-

Table 1. Critical concentration values of Dye-SDS Micelles at .30 °C

	Water*)			Phosphate buffer (0.002 mM dyes)				NaCl (0.002 mM dye)				
	0.002	0.003	0.004	6.2	6.95	9.13	7.93	0.001 M	0.0025 M	0.005 M	0.01 M	0.1 M
cmc millimole/liter	2.56	2.56	2.56	0.8	0.8	0.8	0.8	2.0	1.8	1.0	0.8	0.8

*) Concentration of dyes in millimoles/liter

farbstoffes angewandt. Der Farbstoff tritt in Wechselwirkung mit Natriumdodecylsulfatmizellen, aber nicht mit Cetyltrimethylammoniumbromidmizellen. Die Ergebnisse werden auf der Basis einer induzierten Mizellbildung diskutiert.

References

1) *Goddard, E. D., T. G. Jones*, Trans. Faraday Soc. **49**, 980 (1953).

2) *Colichman, E. L.*, J. Am. Chem. Soc. **72**, 1834 (1950).

3) *Colichman, E. L.*, ibid. **73**, 1795 (1951).

4) *Klevens, H. B.*, Chem. Revs. **47**, 1 (1950).

5) *Klevens, H. B.*, J. Phys. Chem. **58**, 541 (1954).

6) *Mukherjee, P., K. J. Mysells*, J. Am. Chem. Soc. **77**, 2937 (1955).

7) *Hiskey, C. F., T. A. Downey*, J. Phys. Chem. **58**, 835 (1954).

8) *Hartley, G. S.*, Trans. Faraday Soc. **30**, 444 (1934).

9) *Fendler, E. J., J. H. Fendler*, Adv. Phys. Org. Chem. **8**, 271 (1970).

10) *Mukherjee, P., K. J. Mysels*, Natl. Stand. Ref. Data Ser., Natl. Bur. Stand., No. 36 (1971).

Authors' address:

Dr. *H. N. Singh*
Department of Chemistry
Aligarh Muslim University
Aligarh – India

Progr. Colloid & Polymer Sci. **63**, 116–119 (1978)

ISSN 0340-255 X

Osaka Municipal Technical Research Institute, and Department of Applied Chemistry, University of Osaka Prefecture, Osaka (Japan)

Effect of urea on the solubility of benzene and toluene in water

M. Ueda, A. Katayama, N. Kuroki, and *T. Urahata*

With 2 figures and 1 table

(Received December 21, 1976)

Introduction

It is well known that an addition of urea enhances the solubility of hydrocarbons in water. From various experiments it was concluded that urea was a structure breaker and produces structural change in water. However, the mechanism is not better understood than that of the interaction of urea with water.

The purpose of this work is to determine the solubility (N: mole fraction unit) of benzene and toluene in water and urea-water solutions and to obtain a further understanding of the enhancement mechanism of urea.

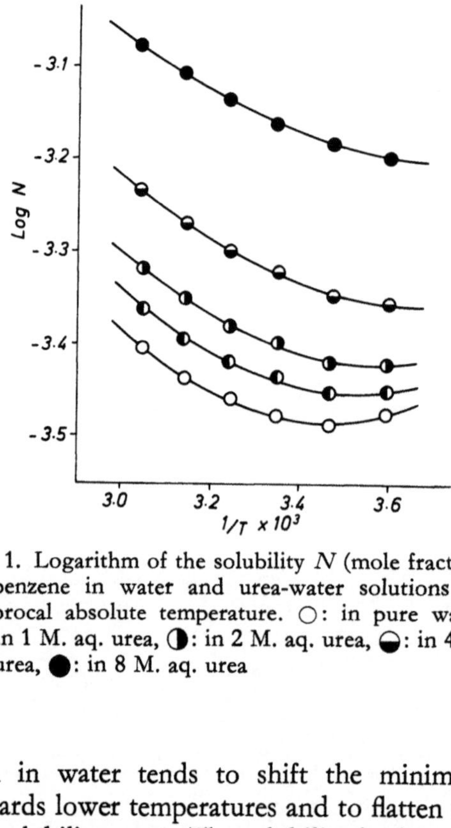

Fig. 1. Logarithm of the solubility N (mole fraction) of benzene in water and urea-water solutions vs. reciprocal absolute temperature. ○: in pure water, ◐: in 1 M. aq. urea, ◑: in 2 M. aq. urea, ◒: in 4 M. aq. urea, ●: in 8 M. aq. urea

Experimental

The solubility of benzene and toluene in water and 1, 2, 4 and 8 mol/l urea-water solutions was determined at 5–10 degrees intervals from the freezing point of each solution to 55 °C. The method of solubility measurement was the same as that described in a previous paper (1). At least two separate runs, and generally more, were made for each sample at above given temperatures.

Results and discussion

The average values were plotted as log N vs. $1/T$ where T is the absolute temperature. The resulting plots are given in figures 1 and 2.

As can be seen from figures 1 and 2, every solubility curve obtained in this work has a almost parabolic form. Further it is noteworthy that the solubility curves for toluene in urea-water solutions, up to 4 mol/l, are very similar in shape to that in water, each having a minimum. The increase in the urea concentration in water tends to shift the minimum towards lower temperatures and to flatten out the solubility curve. The solubility for benzene at temperatures below 5 °C could not be determined, because benzene freezed under these conditions. This seems to be the reason why the solubility curves for benzene in 2, 4 and 8 mol/l urea-water solutions show no minimum.

The solubilities for the benzene hydrocarbons were fit to parabolic equations of the form

$$\log N = A(1/T)^2 + B(1/T) + C. \qquad [1]$$

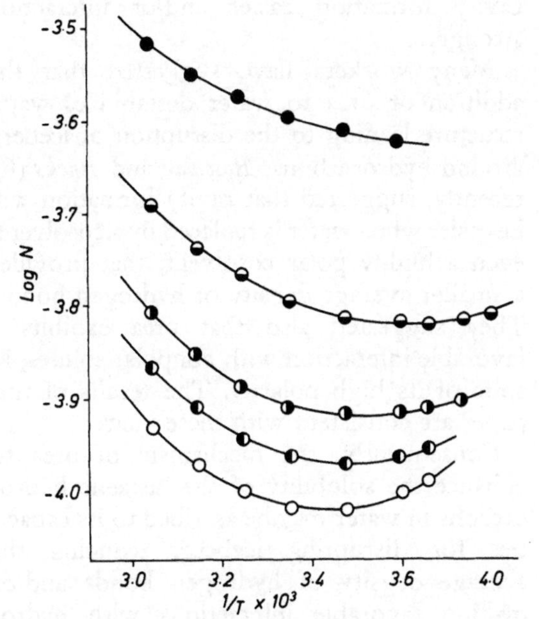

Fig. 2. Logarithm of the solubility N (mole fraction) of toluene in water and urea-water solutions vs. reciprocal absolute temperature. ○: in pure water, ◐: in 1 M. aq. urea, ◑: in 2 M. aq. urea, ◒: in 4 M. aq. urea, ●: in 8 M. aq. urea (below 0 °C urea deposited)

The constants A, B and C for this equation are listed in table 1, which are determined by the method of least squares. The residual-mean-squares between the observed and calculated values is about 5×10^{-6}. This is within the experimental error of about $\pm 0.5\%$ in N.

The constants A and B will characterize the solution process of the benzene hydrocarbons in water and urea-water solutions. This will be discussed later in detail. The constant C is considered to be an imaginary solubility at an infinite temperature.

The heat of solution is proportional to the slope of the curve formed by plotting the logarithm of the solubility versus the reciprocal of the absolute temperature. Therefore, the corresponding expression for the heat of solution, ΔH, can be obtained by differentiating equation [1] with regard to $(1/T)$.

$$\Delta H = -R[2A(1/T) + B] \quad [2]$$

According to equation [2], the heat of solution consists of two terms, that is $-2A(1/T)R$ and $-BR$, where R is the gas constant. The former is dependent on temperature and the latter independent of temperature.

Here, we assume that the observed heat of solution can be expressed as a sum of the heat of iceberg formation, the heat of cavity formation and the heat of interaction.

$$\Delta H_{\text{obs.}} = \Delta H_{\text{ice.}} + \Delta H_{\text{cav.}} + \Delta H_{\text{interac.}} \quad [3]$$

The heat of iceberg formation is the negative heat due to a more ordered structure around the dissolved hydrocarbon molecules. The heat of cavity formation is the positive heat to be required to make a hole in which a hydrocarbon molecule should be accommodated. Here, the heat of cavity formation does not involve the heat due to the structural change of water. The heat of interaction is the negative heat due to the interactions of hydrocarbon molecules with water and urea molecules, in which we include all solvent-solute interactions other than $\Delta H_{\text{ice.}}$ and $\Delta H_{\text{cav.}}$.

As was earlier pointed out by *Bohon* and *Claussen* (2) the number of water molecules in iceberg around a hydrocarbon molecule dissolved in water would decrease sharply with temperature. Therefore, the heat of iceberg formation should decrease rapidly with temperature. Hence this would be temperature dependent and included in the temperature dependent term, $-2A(1/T)R$, in equation [2]. The heat of interaction may be considered constant in the temperature range studied here.

Table 1. Polynomial coefficients in eqn. [1]

Concn. of Urea	Benzene			Toluene		
(mol/l)	A	B	C	A	B	C
0	424554	−2955.82	1.6606	626526	−4300.59	3.3585
1	361857	−2568.56	1.1059	561375	−3909.10	2.8334
2	322365	−2329.40	0.7861	497592	−3506.00	2.2548
4	271120	−2020.00	0.4016	387348	−2804.53	1.2567
8	189131	−1476.60	−0.3351	286743	−2096.18	0.2089

Therefore, it should be included in the temperature independent term, $-BR$. The heat of cavity formation involves the work of bending and disrupting the hydrogen bonds in water and in aqueous urea to make a hole. This may vary with temperature. The raise of temperature would change the size and shape of interstitial cavities in water and in aqueous urea resulting in the change in the number of hydrogen bonds to be bended and broken to make a hole. However, in this paper the heat of cavity formation should be included in the temperature independent term, $-BR$, because the values of $-BR$ obtained in this work are always largely positive. The temperature dependence of the heat of cavity formation may be comparatively small as compared with that of the heat of iceberg formation.

From the above assumptions one can assign $-2A(1/T)R$ to the heat of iceberg formation and $-BR$ to a sum of the heat of cavity formation and the heat of interaction. These assignments provide a simple description of the heat of solution and an useful way of thinking about the solution process of the benzene hydrocarbons in water, even though these are based on no detailed theories.

In order to express the solubility of hydrocarbons in water over a wide range of temperature, several interpolation equations have been proposed by *Alexander* (3), *Arnold* et al. (4) and *Franks* et al. (5). However, neither of them seem to be suitable for a simple description of the heat of solution.

It is difficult or impossible to evaluate the heat of cavity formation and the heat of interaction, separately. The heat of cavity formation, however, would predominate at any temperature studied here, because the values of $-BR$ are always largely positive.

Let us now discuss the effect of urea on the heat of solution of the benzene hydrocarbons. Equation [2] provides a starting point for discussion.

As can been seen from table 1, the addition of urea to water affects both A and B. Both A and B decrease in their absolute values with increasing the urea concentration in aqueous urea. That is, urea makes the heat of iceberg formation more positive and the sum of the heat of cavity formation and the heat of interaction more negative. This indicates that urea makes iceberg formation more difficult, cavity formation easier and/or interaction stronger.

Many workers have suggested that the addition of urea to water destabilizes water structure leading to the disruption of iceberg around hydrocarbons. *Roseman* and *Jencks* (6), recently, suggested that cavity formation will be easier when water is replaced by a cosolvent, even a highly polar cosolvent, that provides a smaller average density of hydrogen bonds. They suggested also that urea exhibits a favorable interaction with nonpolar solutes, in spite of its high polarity. The results of this paper are consistent with these views.

Consequently, the mechanism of urea to enhance the solubility of the benzene hydrocarbons in water may be ascribed to its capacities for disrupting iceberg, reducing the average density of hydrogen bonds and/or making favorable interactions with hydrocarbons.

The expression for the temperature, at which the solubility curve has a minimum, can be obtained easily, if ΔH is taken zero in equation [2].

$$T_{\text{min.}} = -2A/B \qquad [4]$$

As has been pointed out above, $T_{\text{min.}}$ shifts towards lower temperature with increasing the urea concentration in its aqueous solution. This indicates that A decreases more rapidly with the urea concentration than B does.

Summary

The solubility of benzene and toluene in water, and 1, 2, 4 and 8 mol/l urea-water solutions was determined at 5–10 degrees intervals from the freezing point of each solution to 55 °C. The solubilities (N: mole-fraction) were fit to parabolic equations of the form

$$\log N = A(1/T)^2 + B(1/T) + C. \qquad [1]$$

By differentiating the equation [1] with regard to $(1/T)$ the heat of solution was obtained as follows;

$$\Delta H = -R[2A(1/T) + B].$$

The heat of solution consists of two terms, that is, $-2A(1/T)R$ and $-BR$. The coefficients A and B decreased with increasing the urea concentration in its aqueous solution. From the results the mode of action of urea on the solubility of aromatic hydrocarbons in water was ascribed to its capacities for (1) disrupting iceberg, and (2) reducing the average density of hydrogen bonds and/or making favorable interactions with hydrocarbons.

Zusammenfassung

In der Arbeit wird auf die Löslichkeit von Benzol und Toluol in Wasser und wäßrigen Harnstofflösungen in Abhängigkeit von der Temperatur berichtet. Der Einfluß des Harnstoffes auf die Löslichkeit der aromatischen Kohlenwasserstoffe in Wasser wird auf die Störung der Eisbergstruktur, die Erniedrigung der mittleren Dichte der Wasserstoffbrückenbindungen und/oder günstige Wechselwirkungen mit den Kohlenwasserstoffen zurückgeführt.

References

1) *Ueda, M., A. Katayama, N. Kuroki,* Colloid & Polymer Sci. **252**, 486 (1974).

2) *Bohon, R. L., W. F. Claussen,* J. Am. Chem. Soc. **73**, 1571 (1951).

3) *Alexander, D. M.,* J. Phys. Chem. **63**, 1021 (1959).

4) *Arnold, D. S., C. A. Plank, E. E. Erickson, F. P. Pike,* Ind. Eng. Chem., Chem. Eng. Data Ser. **3**, 253 (1958).

5) *Franks, F., M. Gent, H. H. Johnson,* J. C. S. 1963, 2716.

6) *Roseman, M., W. P. Jencks,* J. Am. Chem. Soc. **97**, 631 (1975).

Authors' addresses:

Dr. *Mitsuo Ueda* and Dr. *Toshihiro Urahata*
Osaka Municipal Technical Research Institute
38, Kitaogimachi Kitaku, Osaka 530 (Japan)

Dr. *Akira Katayama* and Dr. *Nobuhiko Kuroki*
Dept. of Applied Chemistry, College of Engineering
University of Osaka Prefecture
Sakai-city, Osaka 591 (Japan)

Für die Schriftleitung verantwortlich: Prof. Dr. F. H. Müller, Marburg-Marbach
und Prof. Dr. A. Weiss, München
Dr. Dietrich Steinkopff Verlag, Saalbaustraße 12, Postfach 111008, 6100 Darmstadt 11
Herstellung: Meister Druck, 3500 Kassel

Mitarbeiter-Bedingungen · Note to Contributors

Originalbeiträge sind an die folgenden Herren zu senden:

Arbeiten aus dem Bereich der Polymerforschung an

Prof. Dr. F. H. Müller (Bereich Polymere Fachbereich Physikal. Chemie, Marburg/Lahn) Haselhecke 26, 3550 Marburg-Marbach;

Arbeiten aus dem Bereich der Kolloidchemie und Biochemie an

Prof. Dr. A. Weiss (Institut für Anorganische Chemie der Universität München) Meiserstr. 1, 8000 München 2;

Die Zeitschrift veröffentlicht nur angeforderte Originalbeiträge zu jeweils einem bestimmten Thema pro Band.

Manuskripte sollen in zweifacher Ausfertigung eingereicht werden. Ihr Eingang wird umgehend bestätigt. Ihr Inhalt muß unveröffentlicht sein. Die Verantwortung für den Inhalt liegt bei den Autoren. Publikationssprachen: Deutsch, Englisch oder Französisch. Jedem Manuskript ist eine Zusammenfassung in deutscher und englischer Sprache beizugeben. Die Typoskripte müssen einseitig und weitzeilig geschrieben sein. Abbildungen sind mit Legenden zu versehen und als klischierfähige Vorlagen einzureichen, wobei die Beschriftung auf einem transparenten Deckblatt anzubringen ist. Formeln bitte deutlich schreiben, insbesondere griechische Buchstaben und Indices! Die Zahl der Abbildungen und Tabellen ist auf das unbedingt Notwendige zu beschränken. Für Literaturangaben gelten die international üblichen Regeln. Die Literatur ist am Schluß der Arbeit zusammenzufassen. – Anstelle eines Honorars erhalten die Autoren insgesamt 75 Sonderdrucke kostenlos, weitere Exemplare auf ausdrücklichen Wunsch gegen Berechnung. – Kosten für nachträgliche Autorkorrekturen, soweit es sich um Textergänzungen in der Druckfahne handelt, werden dem Autor in Rechnung gestellt. – Ausführliche Sonderdrucke der geltenden Mitarbeiterbedingungen sind kostenlos beim Verlag erhältlich. – Nicht den Richtlinien entsprechende Manuskripte werden zurückgesandt.

The authors are requested to submit their **manuscripts** to the following Editors:

Contributions on Polymer Science to Prof. Dr. F. H. Müller;

Contributions on Colloid Science and Biochemistry to Prof. Dr. A. Weiss;

This journal will publish original contributions only on request by the editors covering the special scope of each volume.

Manuscripts should be submitted in duplicate and should contain original work as yet unpublished elsewhere. Their receipt will be acknowledged promptly. Authors are fully responsible for the contents of their contributions. Publications languages: English, French or German. Each manuscript should include a summary in English and German. All manuscripts should be double-spaced, typed on one side only. Illustrations and drawings should be made carefully, with India ink on white drawing paper, blue tracing linen or coordinate paper ruled in blue only. Lettering at the sides of graphs may be pencilled in and will be typeset. Legends must accompany the drawings. Formulas, symbols and Greek letters should be carefully made and annotated and subscripts and superscripts clearly shown. The number of figures and tables should be held to a minimum. The list of references should be written on a separate page. It is recommended that abbreviation of the titles of the Journals be made in conformity with Chemical Abstracts (see List of Periodicals, 1961).
Authors will receive 75 reprints of their contribution free of charge and may order an additional number at cost. Authors making elaborate alterations and additions in proof will be required to bear the costs thereof. More detailed instructions to the authors can be obtained from the publisher free of charge. Manuscripts which do not conform with the above guidelines will be returned to the authors.

Geschäftliche Bedingungen · Note to Subscribers

Erscheinungsweise:

Zwanglos nach Bedarf in Bänden verschiedenen Umfangs.

Bezugspreis dieses Bandes:

DM 85,– plus Porto.
Bezieher der Zeitschrift „Colloid and Polymer Science“ erhalten den Band automatisch im Rahmen ihres Abonnements mit 20 % Nachlaß.

Die Zeitschrift wird automatisch zur Fortsetzung weitergeliefert, sofern nicht vier Wochen vor Jahresende eine Abbestellung vorliegt.

Photokopier-Wertmarken:

Für jedes Photokopierblatt eines Beitrages oder Beitragsteiles aus dieser Zeitschrift ist eine Wertmarke von DM –,40 zu verwenden, erhältlich bei der Inkassostelle für urheberrechtliche Vervielfältigungsgebühren GmbH., 6000 Frankfurt a. M. 1.

Frequency of Publication:

Irregularly in volumes of different size.

Subscription rate of this volume:

DM 85.– plus postage.
Subscribers to "Colloid and Polymer Science" will receive this volume additionally with a 20 % discount.

The subscription will be extended automatically for unless there is a cancellation received four weeks before the end of each year.

Photostat-Stamps:

Each photostat-sheet of an article or part of an article published in this journal must show a stamp of DM –,40, which may be obtained by the Inkassostelle für urheberrechtliche Vervielfältigungsgebühren GmbH., 6000 Frankfurt a. M. 1.

Verlag, Copyright und Anzeigenverwaltung: / **Publisher, Copyright and Advertising Manager:**

Dr. Dietrich Steinkopff Verlag, 6100 Darmstadt, Postfach 1008, Telefon (Phone): (06151) 21766 – Postscheckkonto (Postal Account) Frankfurt a. M. 95697 – Bank (Bankers): Deutsche Bank Darmstadt No. 260117. Foreign subscribers are advised to pay by cheque.

Anzeigenverwaltung: / **Advertising Manager:**

Dr. Karl Niedermeyer Nachf., 6000 Frankfurt a. M.-90, Georg-Speyer-Str. 76, Telefon (Phone): (0611) 775036
Postscheckkonto (Postal Account): Frankfurt a. M. 19383.

Titel-Abkürzung: / **Abbreviation of Title:**

Colloid & Polymer Sci.
Printed in Germany

GPSR Compliance

The European Union's (EU) General Product Safety Regulation (GPSR) is a set of rules that requires consumer products to be safe and our obligations to ensure this.

If you have any concerns about our products, you can contact us on ProductSafety@springernature.com

In case Publisher is established outside the EU, the EU authorized representative is:

Springer Nature Customer Service Center GmbH
Europaplatz 3
69115 Heidelberg, Germany

Batch number: 10578403

Printed by Printforce, the Netherlands